AN ARCHAEOLOGIST'S GUIDE TO ORGANIC RESIDUES IN POTTERY

Archaeology of Food

AN ARCHAEOLOGIST'S GUIDE TO ORGANIC RESIDUES IN POTTERY

ELEANORA A. REBER

The University of Alabama Press Tuscaloosa

The University of Alabama Press
Tuscaloosa, Alabama 35487-0380
uapress.ua.edu

Typeface: Minion

Cover images: Rimsherds from the George Reeves site in Illinois;
photos by Eleanora A. Reber
Cover design: Michele Myatt Quinn

Cataloging-in-Publication data is available from the Library of Congress.
ISBN: 978-0-8173-2122-2
E-ISBN: 978-0-8173-9398-4

To my parents, James and Clarice Reber

CONTENTS

Preface ix

Acknowledgments xi

Introduction: Organic Residues in Pottery and the Archaeologist 1

1 Residue Formation, Composition, and Preservation 8

2 Excavating the Residue: Extraction, Instrumentation, and Analysis 19

3 Residue Contamination: How to Detect and Avoid It 44

4 Alkaloids and Other Mind-Altering Substances 61

5 Plant Resins 86

6 Compound-Specific Isotopic Analysis: Detecting Maize and Isotopically Unique Resources in Residues 105

7 Fish and Shellfish Detection 129

8 Sample Size and Residue Analysis 147

9 The Big Picture 165

Glossary 175

References Cited 187

Index 205

PREFACE

CHEMISTS LARGELY ESTABLISHED and developed the field of archaeological organic residue analysis in pottery, and many of them quickly learned some archaeology over the course of their communication with colleagues. A few archaeologists obtained training in organic and instrumental chemistry and came at the field from the other direction. I am somewhat unusual in that I have been trained as both an archaeologist and a chemist. At Beloit College, I earned a BS in Chemistry and a BA in Anthropology. I obtained my PhD at Harvard University in anthropology by successfully developing a technique for the identification of maize in pottery residues (described later in this book) while also excavating in the American Bottom region of the United States. My subsequent career, spent at the University of North Carolina Wilmington (UNCW), has involved excavations in Illinois and in North Carolina and also a wide range of archaeological pottery residue analysis extending over about 20 years. I am the director of the UNCW Pottery Residue Lab and of the UNCW Summer Field School in the Cape Fear region. My inability to make up my mind about which field to follow means that I approach archaeological problems with a chemist's perspective and chemical problems with a solid grounding in excavation practice. Hopefully this background will be helpful in communicating the complexities and possibilities of archaeological residue analysis to an archaeological audience.

My goal in this book is to produce an introductory guide to organic residue analysis in pottery, as plainly and simply written as possible. It is meant primarily for archaeologists with an interest in residue analysis and for the general reader with some background in archaeology. Many technical terms are therefore not used in the text or are introduced gradually. To help the reader segue into the more technical residue literature, a glossary is included at the end of the book.

As this book aims for a plain and informal style, I use the first-person and second-person point of view. For similar reasons, and to avoid being binary, I use "they/them" as a gender-neutral singular pronoun throughout.

ACKNOWLEDGMENTS

I WOULD LIKE to thank the long list of people who have made this book possible. Wendi Schnaufer, at the University of Alabama Press, has been supportive throughout the publication process, and her patience has dramatically improved your (the reader's) experience. Cindy Mackey at the Peabody Museum of Archaeology and Ethnology at Harvard University was prompt and cheerful in assisting with photo permissions of artifacts from that generous institution. Two anonymous reviewers were kind and helpful during the prepublication process; their suggestions improved the book immeasurably. As usual, all mistakes are my own.

The community of archaeologists in the eastern and midwestern United States and the Caribbean have submitted many residues for analysis over the years and have been patient with ambiguous results and interpretations that have changed and improved through time. John E. Kelly has been particularly forbearing with my inveterate destruction of pottery samples. The community of scholars at UNC Wilmington (UNCW) has been a joy to work with for almost 20 years, helping with everything from emergency loans of instrumentation and lab space to plant identifications to particularly difficult mass spectra. Thank you to Chad Lane, Pam Seaton, Ralph Mead, Ai Ning Loh, Steve Emslie, and many other people. Several generations of students and archaeology classes have helped with residue analysis and background research at UNCW. Some of these have been explicitly named in the text, but many others were not. Thank you all! Finally, my husband, Paul Perkins, and children, James and Julia Perkins, have been supportive throughout the writing process. I could not have written this without them.

INTRODUCTION

Organic Residues in Pottery and the Archaeologist

Every organic residue in pottery is in a way like a separate archaeological feature. Organic residues are made of chemical compounds, or components, that absorb within pots over their use-lifetime. Like features, some residues are more complex than others. Pots that have been used many times to process or cook many different things include many depositional layers of components from these events. Other pots may have been used only a few times, or only to process a few ingredients. These residues are simpler and easier to interpret. Similar to archaeological features, organic residues have issues with diagenesis and taphonomy. Components from food in residues may wash out of the pot wall during rainstorms, be eaten by bacteria, or undergo chemical transformations into other, smaller and simpler chemical compounds. Just as modern intrusions may occur in archaeological features, residues may absorb compounds during burial, excavation, or storage. Compounds from sunscreen, bug spray, labeling, and plastic are the most common of these. Like interpreting a complex feature, interpreting each separate pottery residue can be complicated, and there is considerable room for error. However, like excavating many features on a single archaeological site, the analysis of many pottery residues from a single site allows fascinating interpretations of human behavior that can be reached only in this way.

Organic residue analysis of pottery allows archaeologists to interpret the ways in which people use pottery, connecting theories of form and function in archaeology to the known function of pottery. It allows a more intimate knowledge of how people used pottery on a site: was pottery use generalized, with all vessels being used to process the same wide range of ingredients? Or were different vessels used for different types of ingredients, such as for separate cooking and/or noncooking uses, or for different categories of foodstuffs? Were elite vessels used to process ingredients at all? These sorts of questions, and many more, can be addressed through organic residue analysis.

As with developing a research plan for a complicated archaeological site, these questions can only be addressed if the researchers develop a sampling plan that focuses on the question or questions of interest. Sometimes the question is very simple: Were these pots used to cook or process foods? If so, what? This type of question can be answered with a relatively small number of samples—perhaps 20. More complex research questions require more samples that are more carefully chosen. Very complicated research questions that get at complex aspects of human behavior require large sample sizes of over at least 100 samples, all of which need to be selected to minimize contamination and maximize the information of interest to the excavator.

Organic residue analysis of pottery, like archaeology, is usually more complicated than people think but can produce knowledge about the past that can be obtained only in this way. The goal of this book is to familiarize a general archaeological audience with the possibilities and pitfalls of this type of analysis. An archaeologist does not need to know how to interpret each individual residue, but it is useful to know what sorts of information can be determined and what cannot, and how to tailor research design and sample selection to a particular archaeological question.

To help with this goal, this book provides a variety of case studies that address different problems that archaeologists have wanted to solve over the years, or in some cases problems that showed up naturally as part of ongoing research projects. Like archaeological excavations, residue studies can sometimes produce unexpected results that need further explanation.

A Note on Notes

Throughout the book, brief and specific discussions of issues in the text are presented as boxed *notes. The notes are a sort of bridging mechanism between the case studies and the reminder of the text, offering mini case studies, asides, and slightly peripheral discussions of interesting issues in residue analysis.*

The remainder of the introduction sets the stage for the book as a whole. It includes a quick guide to reading a residue result, although a more detailed version of the same process is explained in chapter 2. Various pitfalls that archaeologists can encounter while trying to apply residue analysis are described as well as a rapid explanation of how to avoid them. It similarly introduces the structure of the book. This allows the reader to skip to chapters of particular interest; however, the book is written so that information from previous chapters will inform later chapters.

HOW TO READ A RESIDUE RESULT (QUICK VERSION)

Every residue is made up of a mixture of many different chemicals. Residue analysis therefore includes a series of steps. First, the residue is extracted from the matrix, either pottery or carbon. Then, the mixture is analyzed in some way that separates the many chemicals from one another, and then identifies them. Finally, the analysis is interpreted. Different types of chromatographic instruments attached to a mass spectrometer are usually used to separate and identify the compounds in a residue, as shown in Figure I.1. Chromatographs separate out complex mixtures, and the mass spectrometer helps identify each compound as it comes off the end of the chromatogram.

Examining a residue analysis therefore involves looking at two different types of figures. The chromatogram shows how much of the different compounds in the mixture there are, and how quickly they were separated. This is a good overview of all the compounds in a residue, each of which appears as a peak in the chromatogram. Each separated compound will then have a mass spectrum, which helps identify it. The easiest way to compare different residues is by looking at their chromatograms. The case studies in this book will include many chromatograms, which will help archaeologists interpret the results of residue analysis in a more informed way.

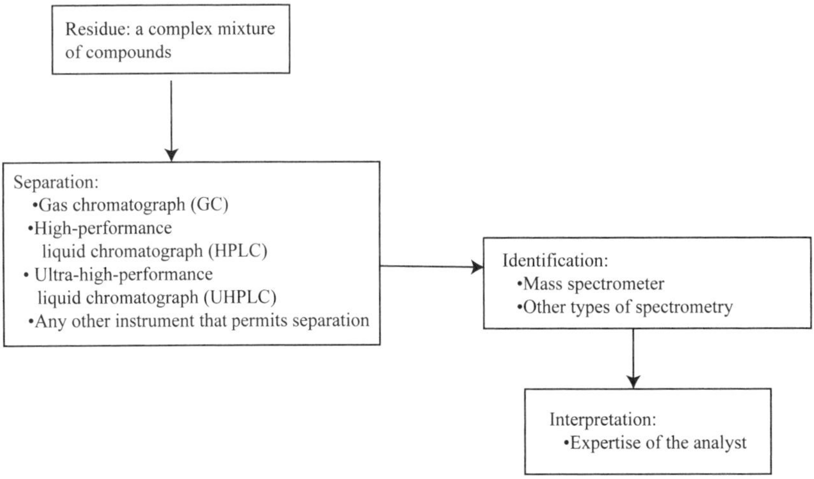

Figure I.1. Simple schematic of instrument setup for the separation and identification of organic residues. (Eleanora A. Reber)

REASONABLE EXPECTATIONS FOR ANALYSIS AND RESULTS

Organic residue analysis in pottery is a fairly recent technical application. It depends on modern chromatographic and spectrometric advances (as described in more detail in chapter 2), which began to be developed during the 1950s. Commercial applications of some of these advances, notably gas chromatography/mass spectrometry (GC/MS) entered general use only in the 1970s. Modern high-performance liquid chromatography/mass spectrometry (HPLC/MS) technology emerged in the late 1980s and is still improving dramatically. The first published use of GC/MS to analyze an archaeological pottery residue dates to 1976 (Condamin et al. 1976) and was used to confirm the presence of fatty acids from olive oil in a Roman amphora. Occasional studies applying chemical techniques to pottery residues took place through the late 1970s and the 1980s. In 1990 Richard P. Evershed and Carl Heron published the first papers on the use of high-temperature GC/MS to analyze mixed organic residues in archaeological pottery (Evershed et al. 1990; Heron and Evershed 1993). Analysis of organic residues in pottery in archaeology therefore functionally began in 1990 and has undergone rapid improvement and development since that time. Because of the technical difficulties involved, and the necessity of understanding organic chemistry in order to interpret the results of the analysis, the field was invented and developed by organic and instrumental chemists, and not by archaeologists. Perhaps because of this, archaeologists have often struggled with understanding reasonable expectations for the analysis and results of residue analysis. There are two relatively common pitfalls that often catch archaeologists off guard with organic residue analysis in pottery.

A Note on Attempting an Analysis with Little Specialized Training

An alert archaeologist might read a paper on organic residue analysis in pottery and note that the technique used is clearly outlined, and that it uses an instrument commonly found on every university campus in the developed world. A chemist interested in collaboration can usually be found, and the two new colleagues begin a research project. Soon, however, it becomes apparent that the archaeological residue extractions do not look like modern reference samples produced for the project. Interpreting the results is unexpectedly difficult, and attempts to publish them are cut to pieces by reviewers who (usually) point out that the interpretation is incorrect and that the study didn't follow established methodologies and precautions. Disappointed, the once-hopeful archaeologist and chemist abandon the field.

Since 1990, an elaborate series of research and laboratory protocols have been developed for archaeological residue analysis. These protocols were developed to minimize contamination during sampling or in the laboratory and to maximize information yield from extracted residues. They are unique to residue analysis and not part of standard chemistry training. In addition, most organic and instrumental chemists are familiar with modern samples rather than complex mixtures that have undergone a series of breakdown processes during burial. It is not particularly easy for a chemist and archaeologist without previous training in residues to take up the study and get reliable results in their first attempts.

The extraction and analysis of organic residues in pottery are relatively simple. The interpretation of these residues is complex and requires a solid knowledge of food chemistry, interpretation of complex mass spectra, and the effects of burial, bacteria, water, oxidation, and reduction on already-complicated mixtures of organic compounds. This sort of interpretive background is specialized and not included in standard chemistry training, which often focuses on synthesis of specific known compounds or the identification of unknown modern compounds. A chemist who is interested in residue analysis is almost certainly going to need some additional training. It is possible to learn this background and enter the field. But like any technical analysis it requires investment of time and effort. Many of the factors that make this field interesting and complicated are outlined in later chapters. But like any technical analysis, it is never wise to assume that the field can be mastered quickly and without serious study. To do so is to minimize the expertise required to perform archaeological residue analysis.

However, another archaeologist might encounter a different pitfall by placing too much confidence in the expertise required to perform this analysis.

A Note on Assuming That Residue Analysis Can Completely Reconstruct Pot Use

Knowing that the field is complex and based on scientific techniques and interpretation that require long training, an archaeologist might assume that the analysis can identify practically any desired resource in an archaeological residue. Identification of particular species of fish or plant in a residue, of the percentage of meat vs. grain cooked in a vessel, or of a previously unknown ingredient, such as beans or nuts, would be extraordinarily useful information for archaeological interpretation. The archaeologist finds a laboratory online or a graduate student at a nearby school who says that they can produce these sorts of results. Soon enough, a report appears that includes fascinating and unique results on the submitted residues. Excited, the archaeologist rushes to publish them. The publication attempts are cut to pieces by reviewers who (usually) point out the lack of any basic research or background on the interpretations, which are at best completely unproven and, at worst, dangerously over-confident and in danger of producing false positive results. Disappointed and embittered, the archaeologist abandons the study and often doesn't believe future results published in the field.

The key to successful use of archaeological pottery residue analysis is a good knowledge on the part of the archaeologist of what sorts of results can be expected and what cannot, and a healthy, but not overconfident, appreciation of the amounts of training required to interpret archaeological residues. Readers of this book will hopefully be ready to avoid both of these pitfalls and apply pottery residue analysis in a productive and useful fashion with reasonable expectations for the types of results that can be obtained.

BOOK STRUCTURE

Chapter 1 explains how residues form and preserve in a sherd. It includes a basic primer of the terminology used to describe compounds in residues and how these compounds are useful in interpretation. The breakdown of residue components during burial on archaeological sites is also included.

Chapter 2 discusses analytical methods of residues, with a comprehensive (and hopefully not too technical) explanation of the ways that GC/MS and HPLC/MS work, how biomarkers are used to identify specific resources in a residue, and how more common but less diagnostic compounds may be used to give a general interpretation of vessel contents. This chapter includes some specific examples of residues and their interpretation. Although wide-ranging predictions are always dangerous in science, this chapter provides broad guidelines on what sorts of resources can be distinctively identified, and which probably cannot be (with suitable caveats).

Chapter 3 addresses the sticky issues of contamination in archaeological residues. Some compounds may wash into a residue during burial. As a particularly dramatic example of this, I discuss the effect of the Deepwater Horizon oil spill and cleanup on pottery residues along the Gulf Coast. Most common and ubiquitous contaminants derive from excavation and storage, however, with sunscreen, bug spray, and plasticizers being the most common. Commonsense guidelines on how to avoid these contaminants are included as well as best practices for when the commonsense guidelines were not followed. The concept of blanks and why they are important will also be explained.

Chapter 4 discusses the detection of psychoactives and alkaloids in pottery residues. Targeted extraction and analysis techniques are particularly useful when detecting many of these unique compounds. Most psychoactive biomarkers are alkaloids such as caffeine and nicotine, but this chapter also discusses the detection of alcoholic beverages in residues. Contamination in the field and lab can be a serious problem with some psychoactive biomarkers, especially caffeine. This chapter describes how to minimize these dangers.

Chapter 5 covers the seemingly ubiquitous presence of tree resins in archaeological residues in North America. Resins are usually chemically stable, distinctive, and unique and as such are among the easiest of resources to identify in residues. Because of the stability and uniqueness of these compounds, however, their interpretation needs input from archaeologists and paleoethnobotanists to determine what the resins mean archaeologically.

Chapter 6 explores the use of compound-specific stable isotope analysis to identify maize and other isotopically distinctive resources. This includes a brief discussion of the chemistry of maize and other grains and a more detailed discussion of how lipids are absorbed within pot walls. This is necessary to explain the phenomenon of "masking" when lipids from very fatty or lipid-rich resources essentially swamp the lipids from lipid-poor resources.

Chapter 7 discusses the identification and interpretation of fish and shellfish. This includes a brief discussion on the lipid contents and chemistry of most fish and shellfish, and of different residue extraction methodologies. Fish and shellfish lipids seem to be unusually liable to staying bound within the ceramic wall during extraction, and so a change in extraction methodology can dramatically affect the detection of these resources in archaeological residues.

Chapter 8 ventures into the fraught and crucial concept of sampling strategies. How many pottery or visible residue samples should be destroyed to answer a particular archaeological question? How much of the sample will be needed? How can variables be minimized in a given set of pottery samples in order to obtain the most useful archaeological results? Each research project is different, so this section will probably not be able to answer every question on this topic. My goal is to include basic guidelines for planning research designs with archaeological meaning and to give a broad idea of how many pottery samples will be needed for different types of archaeological questions.

The final chapter addresses the potential and probable future benefits of archaeological pottery residue analysis. Although interpreting residue analysis can be complex and multilayered, so can the interpretation of archaeological stratigraphy and artifacts. It includes a brief discussion of how to communicate with a residue analyst and the different ways that analytical costs are structured.

1

RESIDUE FORMATION, COMPOSITION, AND PRESERVATION

LIKE AN ARCHAEOLOGICAL site, many processes go into the formation of a pottery residue. Also like an archaeological site, there are several different ways to investigate the residue, depending on the research question of interest. Even more like an archaeological site, interpreting the results can be challenging. A basic understanding of how these issues intersect will clarify the problems and possibilities of residue analysis.

Organic residues in pottery include a wide range of chemical compounds, or components, that absorb within the ceramic matrix of a pot or within a charred layer on the pot and are preserved there throughout burial. As a result, each residue as analyzed reflects a wide range of cooking events and absorption behaviors as well as interactions with soil, water, and bacteria during burial. This chapter discusses the formation of pottery residues from vessel firing through burial and explains how residues form and preserve. It includes a basic primer of the terminology used to describe compounds in residues and how these compounds are useful in interpretation. Finally, it discusses the ways in which residues transform during burial.

Because of the technical nature of chapters 1 and 2, the reader may feel free to skip over them and continue to chapter 3. It will probably be necessary, however, to refer back to these chapters and the glossary to address terminology that appears later in the book.

RESIDUE FORMATION

Organic residues in pottery have two primary forms: absorbed and visible. Visible residues are the obvious carbonized encrustations that appear on vessels. People visualize these when they picture residues, but they are rare at most sites. Absorbed residues are not visible to the excavator, as they are composed of compounds from vessel contents that absorbed within the ceramic wall of the vessel during use. They are more common than visible residues, but also harder to identify during excavation.

A Note on Food vs. Vessel Contents

In general, when analyzing residues, it seems intuitive that we are mostly interpreting cooking. This is the main use of most unglazed pots, and the vast majority of residues are most easily interpreted as deriving from foodstuffs. That said, it's important to keep in mind that cooking is not the only use for ceramic vessels. They can be used for multiple nonfood uses, including production of medicines, resins for waterproofing, dyeing, and such. For this reason, it is wise to use the term "vessel contents" instead of "food," and "use" or "processing" rather than "cooking." It does tend to make residue reports sound persnickety and overly technical, but it also acknowledges that not every pottery vessel is used for food.

Absorbed residues form when compounds from cooking or other uses soak into the unglazed walls of a ceramic vessel as shown in the rough schematic in Figure 1.1. To determine how this might happen, it helps to visualize a pot or frying pan. A cast-iron frying pan is an especially useful image, as cast iron slowly absorbs food compounds, producing the seasoning that makes these pans nonstick.

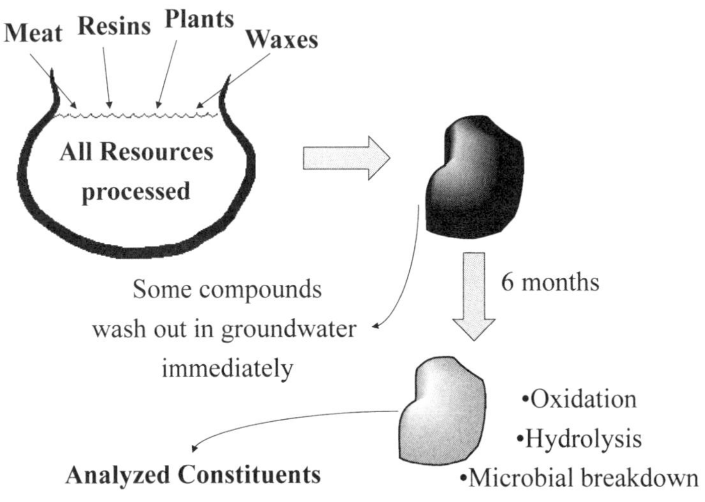

Figure 1.1. Schematic of how absorbed residues form in a pot and how they transform during the burial period. (Eleanora A. Reber)

An unglazed ceramic vessel has been fired enough to vitrify the clay, which is also hot enough to remove all natural soil lipids from the ceramics (Heron et al. 1989; Reber et al. 2019). A freshly fired ceramic pot is therefore empty of organic compounds. When it is first used, some substances from the food or other contents of the vessel will exit the food and get on the ceramic. The process will vary according to the type of cooking used. If the vessel simmered soups, stews, or other contents with water, some constituents of the food will enter the cooking liquid, producing a broth. When you taste a broth, you are tasting a complex mixture of a variety of constituents from the food processed in the cooking pot, including lipids (fats and oils), proteins, and other water-soluble chemicals such as vitamins and flavoring compounds.

If a vessel is used for dry cooking, then fewer compounds will tend to exit the food and get on the pot. Frying meat or other fatty resources will often produce grease (lipids, again) that get on the vessel, but there will be fewer of the other compounds. Again, imagine a frying pan that has seared steaks—it largely contains lipids (the grease) and a very small amount of protein and other compounds. The brown crusty pieces on the pan are often proteins and their breakdown products that have undergone a Maillard reaction, producing more flavor components.

If the ceramic vessel is used as a griddle for starchy foods, such as pancakes, only a few compounds from the food may transfer to the vessel at all, especially if the vessel has been greased to prevent the food from sticking. The residue formed by the cooking episode will be dominated by the grease rather than the actual cooked food. Again, imagine a frying pan after cooking pancakes . . . it is often clean, containing only crumbs or a small residue from the oil used to grease the pan.

Of these potential uses, wet cooking transfers the most food constituents into the cooking fluid. Naturally, only compounds that are water soluble, or that float on the water surface, will be extracted into the broth or stew during cooking. As the vessel used is unglazed, the cooking liquid is partially absorbed within the vessel walls. When there are water-soluble compounds mixed with the cooking liquid, these compounds also absorb within the pot walls. So during this type of cooking, water-soluble compounds are most likely to enter an unglazed pot wall to produce an absorbed residue. Some non-water-soluble compounds that melt out of the food and float on the liquid surface may also absorb within the pot wall in the region of the liquid or cooking line. This is why the neck or any visible cooking line in a pot is a preferred location for residue sampling. When the water evaporates or is emptied from the vessel, the various compounds will remain within the pot wall as a sort of "seasoning layer" similar to that found in a cast-iron frying pan. A similar process occurs with dry-cooking methods, except that fewer compounds will be absorbed within the vessel. With fat frying, any compounds from the fried foods are likely to be overwhelmed by the signature of the frying fat. The longer a vessel is used, the more different compounds absorb within its walls. Types of compounds absorbed will depend on the vessel contents and the cooking method.

It is generally believed that absorbed compounds are protected within small voids in the unglazed ceramic as well as potentially by weak chemical reactions with some of the fired clay components (Craig et al. 2004; Drieu et al. 2019). Because absorbed residues are absorbed within a pot wall, they can form only in unglazed pots. The glaze on glazed pots is basically a layer of glass that prevents food components from reaching the ceramic matrix. The pot will be more waterproof but will not be able to absorb residues.

Visible residues, like the one shown in Figure 1.2, form when some of the carbonaceous material within a vessel scorches during its use. Some compounds from the vessel contents may be preserved or absorbed within vacuoles of the burnt carbon. Since this process does not require water solubility, a wider range of compounds may be found within visible residues than within absorbed residues. Visible residues may also form on almost any surface, including glazed pots, metal, stones, and the like. The only requirement is that some organic substance must be burned and form a carbonation layer.

Figure 1.2. Visible pottery residue deposited on a sherd from Perquimans County, North Carolina. (Eleanora A. Reber)

Both absorbed and visible residues can derive from almost any number of pot uses—from a single episode to a sort of average of many years of uses, all soaking in together into the ceramic wall or carbonized encrustation. One would think that visible residues would tend to result more often from single use (a major cooking disaster), but this isn't always the case. Once a pot scorches, removing the encrustation can be difficult, and some of the encrustation may remain and continue to absorb compounds during use.

Although residues from a single use and from many uses can be easily compared to a single component vs. a multicomponent archaeological site, this is where I must abandon my convenient metaphor. There is no stratigraphy in residue analysis, unlike archaeological sites, so there is no way to determine whether a residue derived from one or many uses. All the lipids in both absorbed and visible residues seem to absorb within their substrate, slowly mixing with other compounds from earlier and later use episodes and migrating within the pot wall (Charters et al. 1993; Evershed 2008a; Evershed et al. 1997). Some residue analysts have suggested that over time voids in the ceramic matrix and vacuoles in carbon encrustations can fill up with compounds from cooking, meaning that residue absorption decreases through time (Heron et al. 1991). Alternatively, Craig and colleagues (2004) suggested that compounds from early vessel uses were carbonized into char through time, so that preserved residues may derive from the most recent uses. There has been no experimental confirmation of either hypothesis, however. It is clear from repeated experimental and analytical work that natural clay lipids are removed from ceramics during firing (Reber et al. 2019), and also that compounds do absorb within ceramic walls and carbon encrustations and preserve fairly well. But the precise physicochemical mechanisms of how this takes place are still undergoing study (Goldenberg et al. 2014; Hammann et al. 2020).

RESIDUE TAPHONOMY: HOW A RESIDUE CHANGES DURING BURIAL

The formation of the residue during vessel use is only the beginning, however. When a vessel stops being used, it will at some point be broken and deposited in the archaeological record. At that point, the compounds in the residue, although protected by voids in the ceramic or vacuoles in the carbon, will continue to age and change through the residue equivalents of taphonomic processes, as illustrated in Figure 1.1. This is why modern experimentally produced residues never look very much like archaeological residues, as shown in Figure 1.3.

When looking at gas chromatograms, as in Figure 1.3, remember that each peak in the graph is a separate compound. The larger the peak, the more of

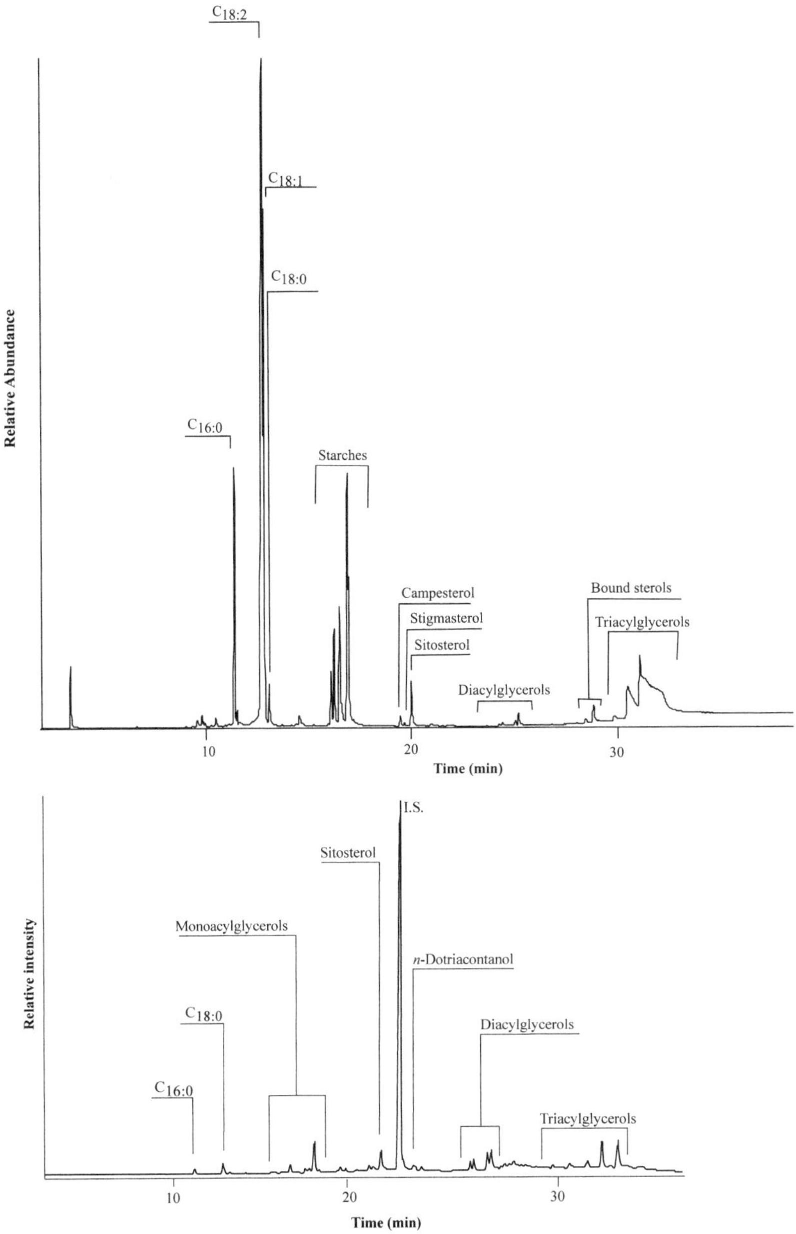

Figure 1.3. Gas chromatogram of lipid extracted from fresh cornmeal (*top*), compared with a chromatogram of cornmeal lipid extracted from a pottery vessel that was buried in clay loam soil for six months (*bottom*). IS = Internal Standard, which has been completely overwhelmed in the fresh extraction. Note the disappearance of starches and bound sterols from the archaeological sherd, and especially the dramatic shift in fatty acid composition between the two extracts. (Eleanora A. Reber)

the compound is present in the sample. To determine about how much residue is present in each sample, most residue analysts add an internal standard, which is a known amount of a particular compound that is unlikely to appear in a natural residue. The sample peaks can then be compared with the internal standard peak to determine absolute amounts of lipids. The variation in the relative amounts of the different peaks in the two chromatograms in Figure 1.3 shows that different compounds are present in each sample in different amounts, even though both derived from cornmeal. The difference in the peak sizes relative to the internal standard peak shows that the overall amount of lipids decreased noticeably after burial.

There are many reasons for these changes. First, as mentioned, most of the compounds that compose residues are water soluble, but many archaeological sites regularly experience precipitation. Compounds that are soluble in cold water tend to wash out of the residue and into the soil. These compounds, often including sugars, salts, vitamins, and light flavoring-type compounds, are lost from the residues. The compounds left after the rainfall are those that are soluble in warm or hot water but insoluble in cold water. The presence of water in an archaeological environment also causes a chemical reaction called hydrolysis in which more complex molecules break down at molecular weak points (usually oxygen bonds) to form smaller compounds. Since smaller compounds tend to be more water soluble, some of these may in turn wash out of the residue and into the soil. Others of these compounds produced by hydrolysis of larger compounds will remain in the residue but will not be in the original form as absorbed within the pot. A similar process happens in the presence of oxygen (oxidation) and bacteria (microbial degradation). Although the ceramic matrix and charcoal encrustations will tend to protect compounds from hydrolysis, oxidation, and microbial degradation, it is not a perfect protection. Compounds with fewer weak points for oxidation and hydrolysis tend to preserve better than those with more of them. This results in better preservation for longer compounds with fewer double bonds and oxygen bonds.

For absorbed residues, the functional result of these taphonomic processes on the residue is that lipids tend to be the dominant component in absorbed residues. Many other compound types either don't absorb within the residue in the first place or are removed from the residue or transformed and removed during burial. This is not completely true—a few nonlipid compounds do preserve in absorbed residues and can produce some remarkably interesting interpretations, most notably alkaloids such as caffeine, nicotine, and scopolomine. That said, most compounds surviving in absorbed residues are lipids. This is not as limiting as it sounds, however. Lipids are a remarkably common, abundant, distinctive class of compound. They include a wide range of different types that can be used to interpret vessel contents. A quick summary of the different types of lipids and the information they provide in residue analysis is shown in Table 1.1.

TABLE 1.1. CHEMICAL STRUCTURES OF LIPIDS AND USE IN INTERPRETING ARCHAEOLOGICAL RESIDUES

Lipid Name/Class and Description	Information Provided
Acyl lipids are formed by a glycerol backbone with a fatty acid attached to an oxygen group on the glycerol. Each glycerol has three possible oxygen linkages. *Triacylglycerols* (TAGs) are intact acyl lipids, with three fatty acids attached to the three oxygens. *Diacylglycerols* (DAGs) have lost a fatty acid to hydrolysis, and only two fatty acids are attached. *Monoacylglycerols* (MAGs) have lost two fatty acids to hydrolysis, and only one remains.	The most common lipid class, found in nearly all fatty lipids. Tends to hydrolyze around oxygen bond linking the fatty acids to the glycerol backbone. The more TAGs in the residue, the better the condition of the lipids.
Fatty acids, described as $C_{16:1}$, in which the subscript first number describes the number of carbons in the fatty acid, and the second number describes the number of double bonds within the chain.	An extraordinarily common lipid type found in nearly all lipid residues, as they are the final breakdown product of acylglycerols. Relative abundance of fatty acids may be used to determine a sort of "average" of contents in a vessel, keeping in mind preservation issues. In general, plant fatty acids are more unsaturated (more double bonds), while mammalian fatty acids are less unsaturated. Fish and shellfish fatty acids tend to be longer in carbon chain and highly unsaturated. These overall compositions are often shifted during burial because double bonds are prone to oxidation. Odd-chain fatty acids often derive from bacteria in some form.
Sterols are multiringed hydrocarbons produced by all complex living organisms.	Cholesterol may be used as a biomarker for fats from mammals, fish, or shellfish. There are also sterols unique to plants (most commonly sitosterol, campesterol, stigmasterol), and fungi (most commonly ergosterol). There are other sterols unique to manuring (most commonly coprosterol) and others that are rare but occur across a range of resources.

TABLE 1.1. CONTINUED

Lipid Name/Class and Description	Information Provided
Waxes are a long-chained mixture of compounds that usually contain fatty acids, alkanols, and sometimes alkanes linked together by an oxygen bond.	Waxes may originate from plants, insects, birds, fish, or mammals. Some waxes contain distinctive amounts of alkanols and alkanes; this is how beeswax is commonly identified. Waxes may hydrolyze to its component parts at the oxygen bond.
Alkanols, described as OL_{16}, in which 16 is the number of carbons in the chain. If there is a double bond, it is noted after a colon. These are long carbon chains with at least one alcohol group.	Usually originate from the hydrolysis of waxes, although they also occur in free form in many lipids. Odd-chain alkanols often derive from bacteria.
Alkanes, described as AL_{16}, in which 16 is the number of carbons in the chain. These are simple long-chain hydrocarbons with no oxygen.	Usually results from the hydrolysis of waxes. AL_{29} is often used as a biomarker for vascular plant waxes (Evershed 2008b). Even-chain alkanes often derive from bacteria.
Terpenoids are wildly diverse compounds made up of small building blocks called isoprenoids.	They are found in all types of living organisms, but primarily in plants (Harborne and Tomas-Barberan 1991). They occur in many shapes and sizes, but the lightest tend to wash out of archaeological residues. Diterpenoids and triterpenoids are often used as biomarkers for specific plant types. These are discussed in greater detail in chapter 5.

In visible residues, a wider range of compounds may survive the residue formation and burial process. This is largely because more water-insoluble compound types may be included in visible residues in the first place. Hydrolysis, oxidation, microbial breakdown, and washing away of water-soluble compounds occur in visible residues as well as absorbed ones.

A Note on Residues in Deserts

Because water is such an important contributor to the removal of compounds from residues and their chemical breakdown, residue preservation is better in arid locations. As I work predominately in the southeastern and midwestern United States, most of the residues I have worked with have undergone extensive water-based transformation during burial. In extremely dry environments, preservation of water-soluble chemicals is noticeably better. A lack of regular hydrolysis also leads to better preservation of complex compounds with oxygen bonds, which includes most lipids, waxes, and carbohydrates (Eerkens 2002, 2005; Serpico and White 1996, and many others). Archaeologists working in a desert environment will find that their yields of residue are more abundant and easier to interpret than those of their colleagues working in rainy environments.

With a basic idea of the variety of lipids and the way we talk about them, it becomes easier to explain how these compounds can break down in the archaeological record. The top portion of Figure 1.3 shows the components of a fresh extract from maize. It shows not just the lipids in Table 1.1 but also a variety of sugars and starches. When maize is cooked in an unglazed pot, most of these components will also absorb within the pot wall. However, when the pot is buried, even for a relatively short time, the residue looks quite different, as shown in the bottom portion of Figure 1.3. The sugars and starches wash out of the pot wall in the North Carolina rain. The same rain causes hydrolysis of the triacylglycerols (TAGs), decreasing the overall amount of these compounds, and increasing diacylglycerols (DAGs) and monoacylglycerols (MAGs). It also increases the number of free fatty acids, but these compounds are themselves prone to oxidation and microbial degradation, and so free fatty acids also decrease in amount and change in composition during burial. Plant waxes present in the maize in small amounts hydrolyze as well, causing an increase in alkanols in the buried residue. Shorter straight-chain compounds, such as shorter fatty acids, alkanols, and alkanes, will tend to either wash out of the residue or undergo microbial degradation to water-soluble fragments that are then washed out of the vessel. Double bonds will tend to oxidize, meaning that the highly unsaturated plant oils typical of fresh maize become more saturated overall (more single bonds and fewer double ones), although still clearly from a plant source. The more double bonds present in a molecule, the more likely they are to oxidize, which tends to bias potential residue interpretation against resources with unsaturated fatty acids, such as fish, shellfish, and plants. This is why longer-chain saturated compounds with fewer oxygen bonds tend to preserve better in residues than shorter-chain, unsaturated compounds with more oxygen bonds.

The complexity of the interactions between residues, sherds, surrounding soil, water, and bacteria is why archaeological residues look so different from a fresh residue extracted from an experimental foodstuff or from a vessel just used to cook an experimental meal. Hydrolysis and oxidation from burial can be partially mimicked by placing experimental residues in soil in a moist oven (Evershed and Charters 1995), but this does not address issues of rainfall or bacteria. It is often simplest when doing experimental comparison to bury sherds for a month or so before analyzing them in order to get at least a quick feel for these reactions. Some experimental work (Reber and Evershed 2004) suggests that most of these reactions take place in the first month or so of burial, so that a relatively short burial will often be a reasonable proxy for taphonomic processes. It's imperfect but is a reasonable compromise between time and completely addressing soil chemistry and microbial effects on residues.

A Note on Soil Chemistry and Residues

Different soil chemistries have different effects on residues. As mentioned, a drier soil will cause less hydrolysis, but there are also clear differences based on particle size (clay vs. sand, etc.) and the microbial suite present in the soil (Reber and Evershed 2004). For this reason, when doing experimental residue production it is often wise to bury the experimental residues in at least two (and sometimes more) different soils as well as keeping a freezer blank in which a residue-containing sherd is kept in the freezer to minimize lipid chemical breakdown for comparison with those buried. Because the soil in the Wilmington area is extremely sandy and often waterlogged (as seen in Figure 1.4), we often send samples to the Piedmont for burial when doing these studies, as the soil is much clayier, drier, and less saline in that region.

Figure 1.4. Sherds from experimentally produced cooking pots about to be buried in a wet, sandy loam soil in Wilmington, North Carolina. A high degree of hydrolysis is likely to occur to the residues absorbed within the sherds. (Eleanora A. Reber)

2

EXCAVATING THE RESIDUE

Extraction, Instrumentation, and Analysis

Once the residue has formed within or on the sherd, and the sherd has been excavated, it is ready for analysis. This is in some ways similar to excavating a single feature on an archaeological site in terms of its information and complexity. In other ways, it is unlike an archaeological excavation. There is no stratigraphy, and the compounds in the residue have to be forcibly removed from their substrate, whether it be a visible residue or an absorbed one.

This chapter gives a quick overview of the ways that residues can be sampled, extracted, and analyzed. It includes a very basic explanation of how the primary analytical methods for residue analysis work and which methods are best for which types of projects and residues. The two primary interpretive pathways for residues are also introduced: the biomarker and fatty acid abundance methods. Finally, this chapter includes a specific example in how to interpret residues, in case study 1. Again, this is a brief, nontechnical introduction. For further subtleties in extraction and instrumentation, it is best to either consult the technical literature or discuss these issues with a residue analyst. I attempt to define all technical terms in the text; however, many of them are also included in the glossary for further clarification if needed.

RESIDUE SAMPLING AND EXTRACTION

When sampling a visible residue, a solvent-washed metal blade of some type is used to carefully remove the charcoal encrustation from the sherd. The charcoal is placed in a solvent-washed vial and then extracted using the selected method. The potsherd itself is unaffected.

When sampling an absorbed residue, a solvent-washed drill of some sort will remove the outer surface of the sherd to minimize modern contaminants such as fingerprint oil and plasticizers from recent processing and storage that may be attached to the pot surface. The analyst then crushes the sherd to a powder. This process destroys the voids that are protecting the residue and maximizes surface area to permit the solvent to dissolve the residue within the sherd. Sample size is generally between 1 and 12 g, depending on the protocol used. Absorbed pottery residue is a destructive analysis, like archaeological excavation itself. At least some of the sampled sherd will be completely reduced to powder.

In the case of very valuable sherds or vessels, burr sampling is a good choice. This method obtains a ceramic sample with minimal destruction by carefully sanding out the interior of the vessel without touching the exterior, as shown in Figure 2.1. This does present some risk, as the sample quantities will be small and the methodology is unable to remove modern

Figure 2.1. Pottery sherd before burr sampling (*top*) and after burr sampling (*bottom*). Damage to the sherd is minimal and is primarily visible as a slight lightening and smoothing of the surface. Standard sampling destroys at least 2 g of the sherd. (Eleanora A. Reber)

contaminants, which can seriously affect interpretation as discussed in chapter 3. Burr sampling is most commonly performed when looking for specific compounds (targeted analysis, as discussed in chapter 4) from museum-quality artifacts.

Types of Extraction and Sampling

Four extraction methodologies are commonly in use for residues. A quick overview of these may help clarify some of the choices involved. The chloroform/methanol extraction method is the oldest and best-tested method, deriving from the original residue studies in the early 1990s (Evershed et al. 1990; Heron and Evershed 1993); it is sometimes referred to as solvent extraction. The residue is extracted from the ceramic or charcoal matrix using a mixture either of 2:1 v/v chloroform/methanol or of 2:1 v/v dichloromethane/methanol. This produces the total lipid extract (TLE), which includes the overall contents of the residue including acyl lipids. Part of the extracted TLE is then reacted with alkali to split the acyl lipids apart and produces two additional fractions. The neutral fraction holds the separated neutral compounds such as sterols, alkanols, and alkanes, and the fatty acid fraction includes the separated fatty acids from all acyl lipids (Charters et al. 1993a; Dudd et al. 1999). It allows evaluation of the overall quality of the residue by looking at the condition of acyl lipids in the TLE. Because of its long-term popularity, many studies in residue analysis were produced using this method.

Recently, two methods designed to optimize fatty acid extraction and speed of laboratory analysis have entered common use. In the alkaline methanol extraction method, the residue is extracted with alkaline methanol, skipping the TLE step and producing only neutral and fatty acid fractions (Craig et al. 2004). It dramatically improves the extraction of fatty acids, uses small sample sizes, and requires two instrument runs instead of three. It does not permit evaluation of the overall quality of acyl lipids in the residue, and so information about acyl lipid preservation is lost (Craig et al. 2004).

In the acidic methanol extraction method, the residue is extracted with acidic methanol, producing a single fraction that includes all fatty acids and that may be adjusted to include all neutral lipids as well (Correa-Ascencio and Evershed 2014). Similar to the alkaline extraction method, it does not permit evaluation of the overall quality of acyl lipids in the residue but has better extraction of fatty acids and uses a small sample size of 2 g. It also requires one instrument run instead of two or three (Correa-Ascencio and Evershed 2014). Recent studies suggest that acidic methanol extraction of neutral compounds is better than or equivalent to chloroform/methanol extraction of these compounds, though there is some variability probably based on compound type, the physicochemical structure of the clay, and the overall composition of the residue (Reber 2021).

Researchers interested in water-soluble alkaloids such as caffeine often use a water or hot water extraction (Hurst et al. 1989; Washburn et al. 2014, and many others). This method produces superior results for the extraction of caffeine and other water-soluble alkaloids, discussed in chapter 4. Water extraction does not work on lipid extractions, however, as most lipids are not particularly water soluble. Double extraction methods, for extracting both lipids and water-soluble compounds, are discussed in chapter 4.

A Note on the Bound Insoluble Fraction

Extraction does not remove all the absorbed residue from a potsherd, especially when using the chloroform/methanol extraction method. A number of studies have established that some residue compounds remain bound to the ceramic after sampling (Craig et al. 2004; Goldenberg et al. 2014; Hammann et al. 2020; Regert et al. 1998; Stern et al. 2000). This residue remaining in the ceramic powder is often called the bound insoluble fraction, and it is a topic of great interest in residue analysis. Acidic and alkaline extraction methods are tailored to remove more compounds from the ceramic than chloroform/methanol extraction, leaving less bound insoluble fraction behind. These methods allow a more complete interpretation of residues with less sample destruction because residue removal is more efficient. In addition, an understanding of the types of compounds left in the bound insoluble fraction gives insight to the ways in which compounds are preserved in absorbed residues. More highly branched compounds, such as isoprenoid fatty acids, appear to be retained more tightly within the ceramic and are not easily removed using neutral solvent, such as chloroform/methanol (Correa-Ascencio and Evershed 2014). This explains why these compounds have been more commonly extracted since acidic and alkaline extraction methods gained in popularity, leading to an increased incidence in positive identification of fish and shellfish in recent years, as discussed in chapter 7.

Residue Extraction and Analysis

When the selected solvent mixture is added to the powdered sherd or charred residue, the mixture is stirred to increase the extraction of residue from the substrate into the solvent, through either ultrasonication or whirlimixing. This mixture is then separated by centrifuge, and the solvent is removed from the solids. At this point, the solvent will undergo further processing to remove and purify the lipids, depending on the type of extraction chosen. Following this step, the residue is blown down or reduced to the

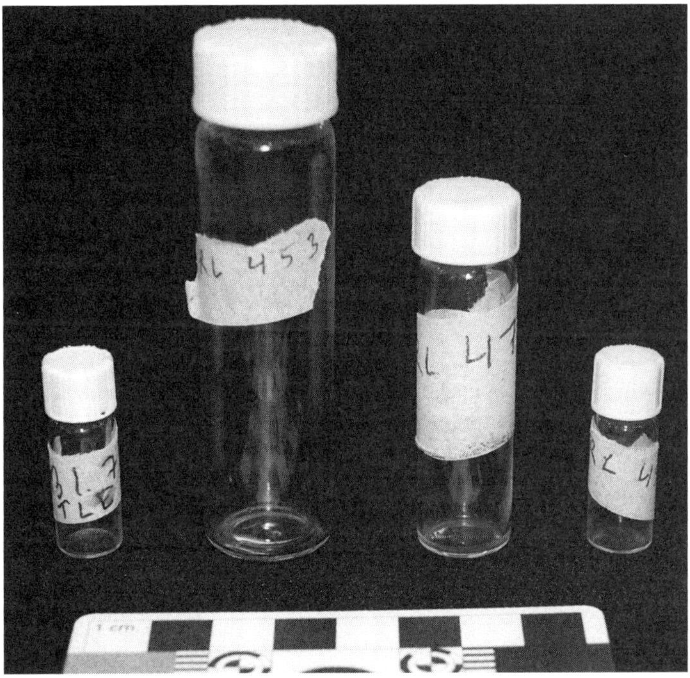

Figure 2.2. Residues extracted from (*left to right*): 1-g blue catfish flesh, an archaeological potsherd using acidic methanol extraction, and two archaeological potsherds using chloroform/methanol extraction. The residue is the cloudy smear at the bottom of the vials. Very abundant residues sometimes appear yellow. (Eleanora A. Reber)

pure lipid. The result, as shown in Figure 2.2, is a vial containing a very small amount of residue.

At this point, various types of chemical analysis are used to separate and identify the mixture of compounds in the extracted residue. As can be seen in Table 2.1, a very large number of different analytical techniques have been applied to residues, both absorbed and visible. All have strengths and weaknesses, although the vast majority of residue analysis now uses the different varieties of chromatographic analyses, as mentioned in this book's introduction. The reason for this should be clear after a brief discussion, and with reference to Table 2.1.

The primary factor in choosing which instrument to use in residue analysis is the common presence of complex mixtures of different compounds in residues. Most chemical analytical techniques are extremely effective at identifying compounds, but only in their pure form. The potential disadvantages of this when doing residue analysis are obvious and become more so

Table 2.1. Types of Analytical Instruments Used in Residue Analysis, with Strengths and Weaknesses

Analytical Instrument	Strengths and Weaknesses
Infrared spectrometry (IR) or Fourier transform infrared (FTIR) spectrometry	Identifies compounds based on molecular bonds' vibrations in the presence of different wavelengths of infrared light. Does not separate mixtures and looks at all bonds in a residue at once. Best for unmixed resources. Strongly affected by potential modern contamination
Nuclear magnetic resonance (NMR) spectrometry	Identifies atoms and molecular structures based on vibrations in the presence of different levels of magnetic fields. Does not separate mixtures. Very precise when identifying pure compounds, but strongly affected by mixture and potential modern contamination
Ultraviolet spectrometry (UV)	Identifies some molecular structures based on vibration in the presence of ultraviolet light. Does not separate mixtures and is strongly affected by solvent choice
Ramen spectrometry	Identifies molecular structures based on vibration in the presence of a laser in a known light range. Does not separate mixtures and is very sensitive to opacity and polarity
Mass spectrometry (MS)	Identifies molecular structure by fragmenting molecules using a source of electronic or chemical ionization and then weighing the different fragments. Does not separate mixtures but can easily be combined with a method that does
Gas chromatography (GC)	Separates complex mixtures that can evaporate at temperatures up to 300°C and then passes the separated mixture on to a detector. Strongly preferred for complex mixtures of largely volatile compounds
Liquid chromatography (LC) or high-performance liquid chromatography (HPLC)	Separates complex mixtures that do not easily evaporate and then passes the separated mixture on to a detector. Strongly preferred for mixtures including compounds that do not evaporate easily and that undergo reactions when heated or for the detection of a small number of target compounds in a residue

given the possibility of false positive results with these techniques. Fourier transform infrared (FTIR) is a good example of this danger. The technique is cheap, available at nearly every chemistry department in the country, and is good at identifying molecular structures. It is often used on pressed pellets but may also be used on liquid residues in various forms. However, the instrument cannot separate complex mixtures. FTIR analysis of a residue will give a sort of hybrid result for all the molecules present in the residue. If the residue derives largely from one resource, the analysis might succeed (Badler et al. 1990). In other cases, this hybrid result mimics a single resource that was probably not present in the original residue. Or, if the residue was contaminated by plasticizers or other modern contaminants discussed further on, the archaeological compounds will essentially be overwhelmed in the FTIR analysis by contamination (Mazow et al. 2014). In addition, some FTIR computer systems attempt a match between the spectrographic results and known modern resources. Because of the mixture of compounds in the residue, and the taphonomic effects of burial on residues, discussed earlier, these matches are extremely inaccurate and sometimes misleading.

The techniques in Table 2.1 have been used on various types of residues at different times, but in most cases they were abandoned relatively early, at least on mixed residues. As briefly described in the introduction, absorbed and visible residue analysis requires some sort of methodology that separates out the many compounds present in the residue, so that each one can be analyzed separately.

CHROMATOGRAPHY AND HOW IT WORKS

Fortunately, there are two major techniques that allow the separation of such complex mixtures: gas chromatography (GC) and liquid chromatography (LC). It's important to note that these techniques do not identify compounds—they are separation techniques. In both cases, a carrier, called the mobile phase, moves the residue through a long, very thin column, called the capillary column. The column is coated on the inside with a thin layer that will interact with different chemicals at different speeds, called the stationary phase. The result is the separation of all the different chemicals in the mixture into separate groupings or waves within the column (called peaks in chromatography). The longer and thinner the column used, the better the separation, but the longer the instrumental run and the more expensive the column. Choice of interior column coating is important but will generally be preselected by the laboratory director based on the standard needs of the lab. The difference between GC and LC is in the mobile phase that moves the residue through the column and the volatility (ability to evaporate easily) of the compounds in the residue.

Chromatography

Figure 2.3 shows a schematic of GC and the way in which separation is largely controlled by evaporation temperature. This technique uses a carrier gas as a mobile phase, with helium usually preferred due to its inertness. The speed and separation of the residue are controlled via temperature variation. The residue components moving through the column will need to be in gas form in order to separate at least most of the time, and heating the chromatographic column causes evaporation of the different compounds in the residue at varying temperatures based on the boiling point of each different compound, as illustrated in Figure 2.3. A gas chromatograph is basically an injection port attached to the long, thin chromatographic column, which is contained in an oven. The column is connected to a detector, which varies based on the needs of the analysis. As will be described, mass spectrometry (MS) is the standard detector for residue analysis. To help separate compounds like acyl lipids and sterols that evaporate at a high temperature, a residue is often treated with a derivatizing agent that makes these compounds evaporate at lower temperatures.

The residue, diluted in solvent, is injected into the injector port with a clean syringe in very small amounts: 1–4 microliters (μL). In the case of on-column injection, the residue is injected directly into the chromatographic column in the oven. In the case of a split/splitless injection, the residue is injected into a glass or metal injection liner that is attached to the column, which is kept heated to a high temperature. The injected compounds rapidly

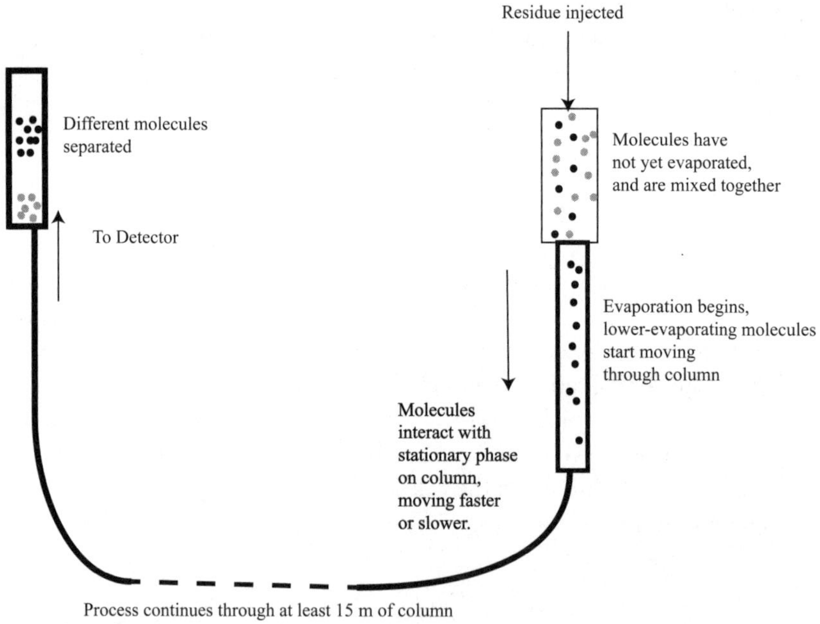

Figure 2.3. Simple schematic showing how gas chromatography separates mixtures of compounds. (Eleanora A. Reber)

evaporate and are deposited on the column at a much cooler temperature, where the previously evaporated compounds are redeposited in liquid form at the beginning of the column. This process may lose more compounds but protects the column from small amounts of dirt or nonevaporating compounds that might otherwise build up on the front end of the column. The injector liner is replaced as needed to keep the column clean.

As the oven slowly heats the column, the different compounds in the column evaporate and flow through the column with the carrier gas. The evaporation temperature thus acts like a starting line for each different compound. As they flow through the column, each compound interacts with the interior coating (stationary phase) of the column, helping to separate compounds with similar boiling points. The idea is that by the time they reach the end of the column, all (or most) of the different compounds will have separated from each other into different peaks that flow into the detector at different times, as seen in Figure 2.3. The different variables in GC are injector type, length and width of column (which affect the speed of the different compounds through the column), carrier gas, carrier gas speed, stationary phase type, and heating program on the column. These variables are complex but are generally set for a given type of residue analysis after the period of stressful and hectic method development that tends to occur when any lab is set up. A GC/MS system can easily fit on a benchtop and includes the GC, the MS, and an attached computer for analysis, as shown in Figure 2.4.

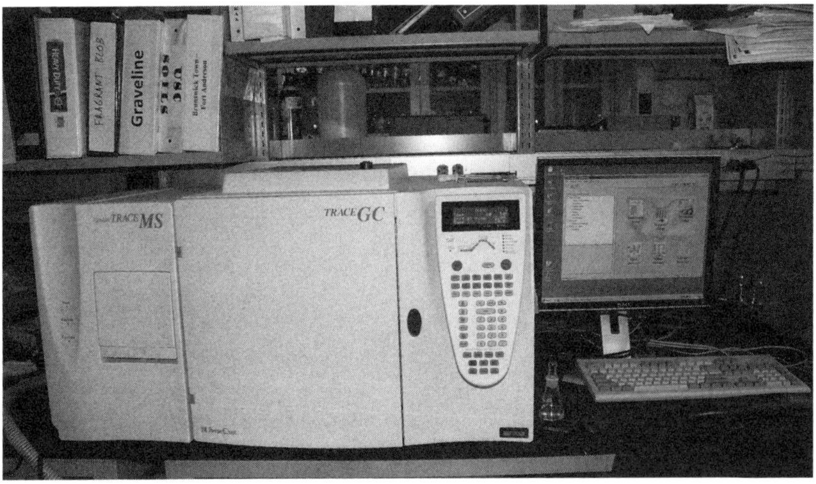

Figure 2.4. Thermo Trace GC/MS system with a manual split/splitless injector. The GC is the largest portion of the instrument, as it must hold the column in an oven. The MS is attached to the left. Both modules feed data to the computer for analysis and interpretation. (Eleanora A. Reber)

In LC, the mobile phase is a liquid, and its composition controls the speed and separation of the residue components. This mobile phase transitions between an *A* solvent and a *B* solvent. For example, an LC sample will be injected into the instrument, and the initial carrier will be 100% *A* solvent. For example, in the standard methodology for caffeine extraction (Washburn et al. 2014), the *A* solvent will be water and 0.1% formic acid. After a brief hold with the water, increasing amounts of the *B* solvent will be added to the column, while the *A* solvent will be decreased. For caffeine, the *B* solvent will be 98.2% acetonitrile and water, plus 0.1% formic acid (Washburn et al. 2014). After a given amount of time, the carrier will be 100% of the *B* solvent. As compounds in the residue dissolve in the mobile phase, they begin to flow through the column. In this case, the starting line occurs when a compound dissolves in the carrier solvent mixture. As each compound moves through the column, it interacts with the stationary phase to help separate different compounds with similar solubilities, similar to the process in GC. Again, like GC, by the end of the column the mixture will hopefully be separated into different peaks that flow through the detector at different times. High-performance liquid chromatography (HPLC) is a variant of LC in which the system is pressurized and solvents are pumped through the column, producing faster run times and better compound separation. It is preferred for most residue applications.

The variables in LC are length and width of column, choice of carrier solvents, speed of carrier solvents, stationary phase type, and the speed with which the two solvents are mixed together during the run. The most important variable in LC is choice of carrier solvents for the mobile phase, which must dissolve the compounds of interest. Conversely, in GC the carrier gas is always the same, and the most important variables are oven temperature and heating speed. An HPLC/MS is similar in size and computing requirements to a GC/MS.

Which One to Use?

The key takeaway from these two techniques is that GC works best on compounds that evaporate below about 300°C, at least when derivatized. Liquid chromatography works best on compounds that don't evaporate at these temperatures. In a very complex mixture like most residues, it is more likely that most of the compounds will evaporate below 300°C than that most of the compounds will dissolve in one of two solvents. As a result, GC is the technique of choice for general residue interpretation. If no one is sure what is in the residue, GC is recommended. HPLC is the technique of choice when looking for certain target compounds, particularly those that do not evaporate easily at temperatures below 300°C, or that dissolve better in water than many other substances. This does not sound terribly helpful, until one considers the characteristics of certain important target compounds. The boiling point of atropine, a biomarker for datura, is 430°C. Theobromine, a methylxanthine found in large amounts in chocolate, and smaller amounts in tea, vaporizes at 290–295°C, but is insoluble in most solvents except for water.

In addition, when an analyst is looking for a small number of target compounds, HPLC is usually preferred, as the solvents may be chosen deliberately to optimize the solubility of the target compounds. With careful solvent choice and use of derivatizing agents, a knowledgeable analyst may use GC to detect these compounds, but most of the high-sensitivity residue analyses on target compounds have used HPLC separation.

A Note on Judging Chromatograms

A good chromatogram (GC or LC) has successfully separated the components in your residue mixture into clear, well-defined peaks. The peaks in the chromatogram should be separate and symmetrical, with pointed tops. If they're unseparated, the detector will measure two compounds at once, making them hard to identify. If they're not pointed at the top, too much of the compound was injected into the instrument, making the quantification unreliable. Quantification helps determine how much of each compound is present in the residue, and unreliable quantification leads to unreliable interpretation, although presence/absence of a diagnostic compound can still be a strong basis for some interpretation. If the peaks are asymmetrical, either a little too much of the compound was injected, the column is aging, or there is a slight problem with the method used. This won't affect interpretation, but it will tend to shorten the life of the column and/or lead to less sensitive results. A good chromatogram should also have a flat, level baseline—that is, the section of the chromatogram with no peaks, or the floor of the graph. Baselines that tilt indicate different types of column problems. There may be a leak, some contamination, or the column may be getting old—the analyst will need to do maintenance or change out the column soon. Many residue gas chromatograms show a slightly higher baseline at the end of the graph, but are otherwise flat. This is because many residue analysts heat their GC columns above 300°C to ensure that all acyl lipids evaporate and leave the column. At these temperatures, the stationary phase in the GC column begins to very slowly break down, raising the baseline. This is why residue analysts tend to run through columns quickly, but it is normal. Baselines that are grassy, with many small peaks, indicate contamination, an aging column, or that there were not many lipids in the residue and the analyst was forced to enlarge it dramatically to see the desired compounds.

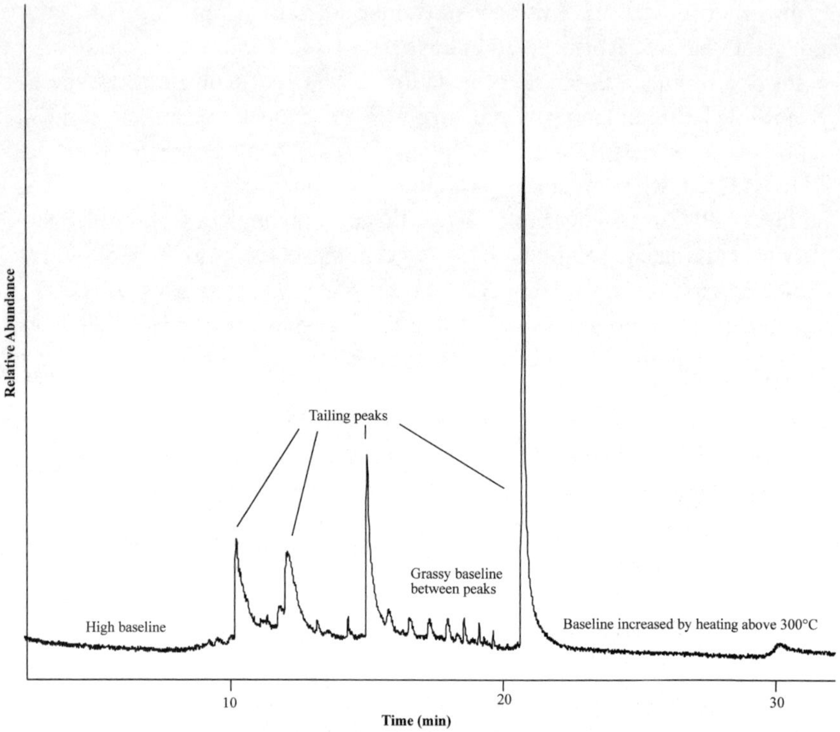

Figure 2.5. Example of a problematic GC. Observe the high and jagged baseline and asymmetric tailing peak shapes. This GC was caused by a combination of an injector leak, an old column, and a residue with a low lipid density. (Eleanora A. Reber)

> *Figure 2.5 gives an example of a problematic chromatogram and Figure 2.6 an example of a good chromatogram. Archaeologists generally do not need to worry about judging chromatograms too often; however, if problematic chromatograms appear in a report, it is useful to know that there might be an instrumental issue.*

MASS SPECTROMETRY AND HOW IT WORKS

Chromatography, again, is a separation technique and not a detection technique. A detector of some sort is usually attached to the end of the column to produce a hybrid instrument that separates complex mixtures and then quantifies or identifies them. The most effective of these hybrid instruments for residues involve mass spectrometry. Mass spectrometry linked to a GC is called GC/MS, and to a LC is called LC/MS or HPLC/MS. In Figure 2.4, the mass spectrometer is the smaller chamber on the left of the figure.

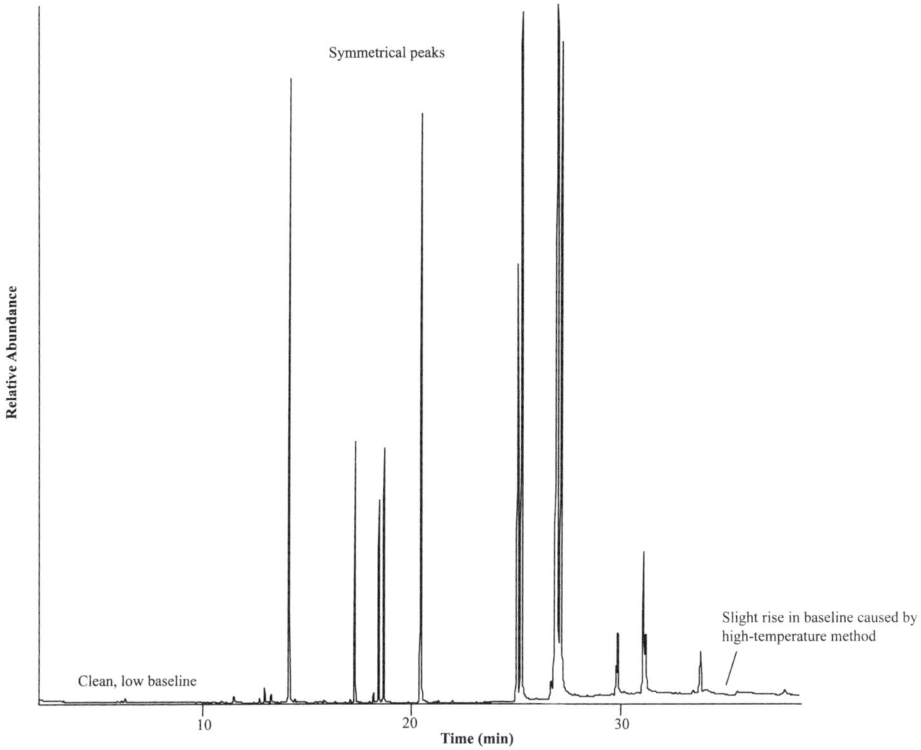

Figure 2.6. Example of a good GC. Observe the low, flat baseline and symmetric, sharp peaks. (Eleanora A. Reber)

Mass spectrometry is a technique that stresses molecules in some way, electronic or chemical, causing them to break down at their chemical weak points. The result is a mixture of fragments from a single compound (separated by GC or HPLC), which can be weighed in the mass spectral detector. The weights of the different fragments are essentially a sort of map to the different parts of the original structure and the places where it is prone to fragment. As these are different in every molecule, the result is a chemical fingerprint. As seen in Figure 2.7, the original molecule can be identified from these fragments.

A search program will attempt a match between the unknown spectrum and entries in a database of known compounds present in the computer attached to the GC/MS or HPLC/MS. In the case of common compounds, such as the fatty acid shown in Figure 2.7, the computer system will successfully match the unknown compound from the residue with a known compound

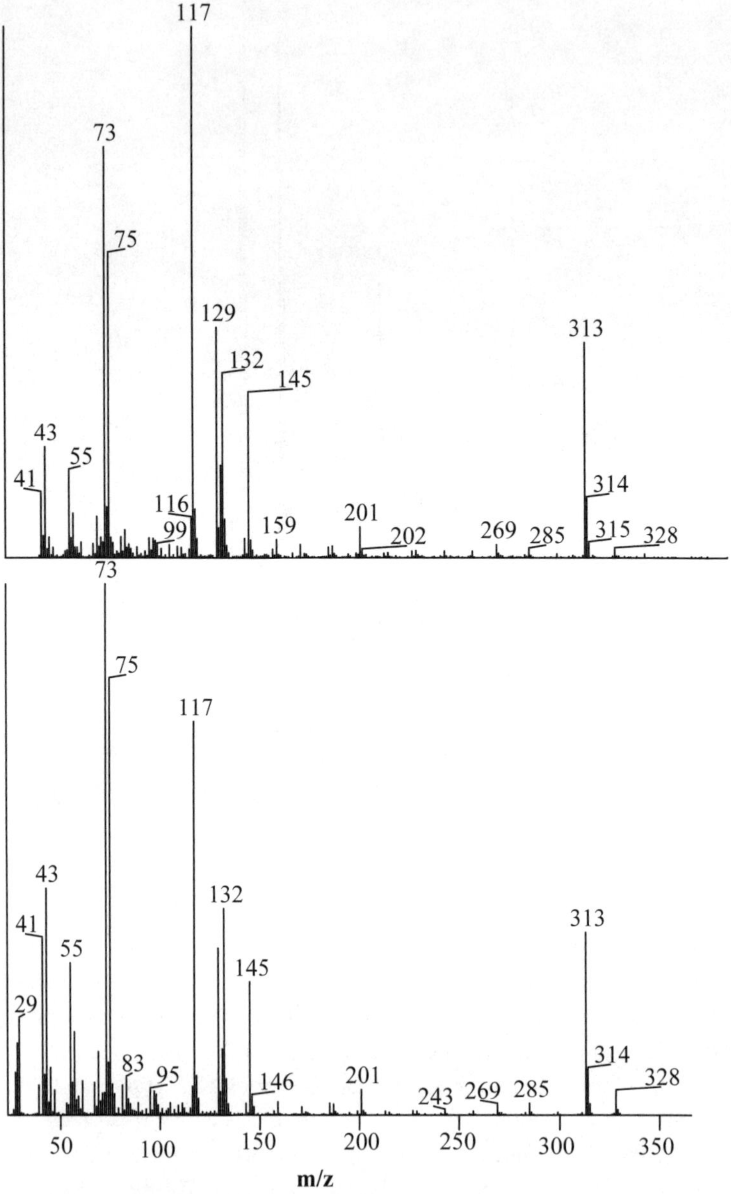

Figure 2.7. *Top*: mass spectrum of an unknown compound from an archaeological fatty acid fraction. *Bottom*: database-assisted mass spectrum match for the common fatty acid $C_{16:0}$, with a confidence of 95.4%. Each vertical line on the mass spectra shows a particular molecular weight, or mass-to-charge ratio (m/z) with the height of the line reflecting the abundance of that fragment. (Eleanora A. Reber)

from the database. A percentage match or confidence interval is always included as well as the MS of the suggested match so the analyst can evaluate the quality of the match.

This procedure is easy, and it's very tempting to believe that interpretation of residues can be essentially automated. The problem lies in the number of unusual compounds in archaeological residues. There are many millions of natural chemicals extant in the world, but only 276,376 electron ionization (the method commonly used by residue analysts) entries in the largest mass spectral database, the National Institute of Technology Mass 17 Spectral database (National Institute of Standards 2014). This database, naturally enough, focuses on relatively common compounds and those that are important in a range of chemical analyses. Although common resources do match easily within the database, such as the fatty acid in Figure 2.7, almost every analyzed archaeological residue will contain multiple compounds not included in the database. To make matters more complicated, the computer system always attempts a match, regardless of whether one exists within the standard database. One of the tasks of the analyst is to evaluate the computer system's match and determine whether it is correct, a match with a similar compound that isn't quite right, or completely off base. If the match is invalid, then the analyst must attempt to reconstruct the molecular structure of the unknown from a familiarity with the literature, with other residues, and from chemical reconstruction of the mass spectrum. This is done by reverse engineering the molecular fragments in the MS. In some cases, the unknown molecule cannot be reconstructed and will remain unidentified. Most residue analysts keep a file of such unknowns and may be able to achieve a match several years after the original project ended.

There are many different types of mass spectrometers that may be attached to a GC or HPLC. The most common is the single quadrupole MS (MSD), which is what most residue analysts use most of the time. For targeted applications requiring more sensitivity, as discussed in chapter 4, one of the other, more expensive MS types might be helpful, such as MS/MS or time-of-flight (TOF). For standard residue analysis, however, MSD works well.

A Note on Reading Mass Spectra

The graphs shown in Figure 2.7 are essential in identifying unknown compounds in residues. In these, the X axis describes the mass-to-charge ratio (m/z) of each molecular fragment, which is the same as its weight in atomic mass units (amu). The Y axis describes the relative intensity. The tallest peak is always 100. Interpretation takes place through a combination of pattern recognition, chemical experience, and practice. The procedure is best described in McLafferty and Tureček (1993). For an *example, Figure 2.7 shows the mass spectrum of $C_{16:0}$, derivatized to a trimethylsilyl ester (TMS ester) for a lower evaporation point. Figure 2.8 shows how the molecule broke apart in the mass spectrometer to produce that spectrum. The peak at 73 is from the TMS ester (28 amu from the silicon, then 15 for each $-CH_3$ group), and 117 is the TMS ester plus the carboxylic acid group, demonstrating that the compound is a TMS ester of a fatty acid. The large peak at 313 reflects the loss of a $-CH_3$ group from the TMS ester, which happens commonly. The total molecular weight is 328. Since the molecule is not stable when heated, the peak at 328 is small. The peaks at 132 and 129 are McLafferty rearrangements. These are reactions that occur within the molecule during the extreme heat and stress of the mass spectrometer. This knowledge is not necessary for a solid understanding of residue analysis, but it may help understand some of the difficulties of residue analysis to realize that the analyst performs these calculations on every peak in every individual chromatogram in a project.*

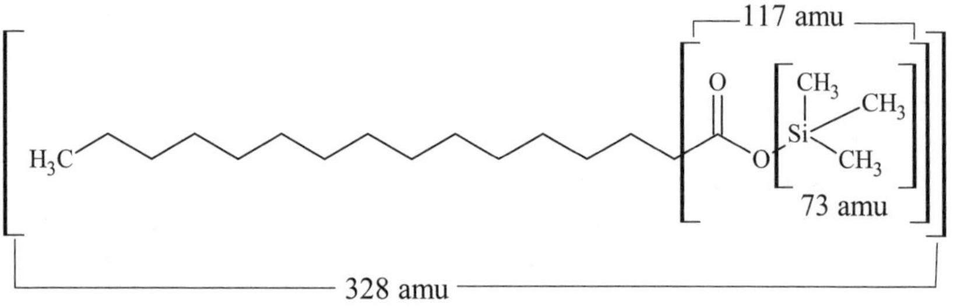

Figure 2.8. How a $C_{16:0}$ fatty acid breaks apart to produce the mass spectrum seen in Figure 2.7, with the molecular weights shown for each bracketed fragment. The loss of a $-CH_3$ group from anywhere on the molecule produces the fragment at 313. (Eleanora A. Reber)

INTERPRETING THE SEPARATED RESIDUE

The combination of complex mixtures and compound breakdown found in archaeological residues is one of the primary reasons that chemists without experience in residue analysis or geochemistry may struggle with residue analysis. Most modern chemists work with single compounds and use MS to determine the success of their synthesis or extraction. Most food chemists are familiar with complex mixtures and the identification of unknown molecules, but not with the results of residue taphonomy and breakdown already described, at least in the quantities found in archaeological residues. Organic geochemists have the body of knowledge most like residue analysts but tend to focus on compounds found in geological resources such as soils and petroleum oils rather than food-derived compounds. The interpretation of residue analysis is therefore a unique subfield of archaeology and chemistry, requiring a different body of mass spectral interpretation than all other fields. This body of knowledge can be mastered, but it takes time, patience, and some background in organic and instrumental chemistry.

Without stratigraphy, as described, all the different surviving compounds in the residue will migrate through the sherd, blending together. Some components in the residue, like fatty acids or alkanes, are very common and may have been present in every resource ever processed in the vessel. For example, the three most common fatty acids in archaeological residues are $C_{16:0}$, $C_{18:0}$, and $C_{18:1}$. They are present in most foodstuffs and many nonfoodstuffs processed in pots. The residue extracted from a visible or absorbed residue will include fatty acids probably from every resource ever processed in the vessel mixed together proportionally according to vessel use. Since this proportion is unknown, fatty acids are difficult (though not impossible) compounds to use when interpreting a mixed residue. When they are used, they are most effective in determining broad resource types that made up the bulk of the components processed in a vessel, as described later in the chapter. They cannot be relied on for more specific identifications, especially in mixed residues (Reber and Evershed 2004; Taché et al. 2019).

The Biomarker Approach

Biomarkers are compounds unique to either specific resources or classes of resources. These are among the most useful compounds in residue interpretation. There are different types of biomarker, all of which provide slightly different information with different degrees of specificity (Evershed 1993, 2008b). Below is a quick summary of the types of biomarkers. More in-depth discussions of many of these will take place in the relevant chapter, later in the book.

Category biomarkers are the most common type. These indicate an overall category of resource processed in the vessel. Sterols, such as cholesterol or sitosterol, are good example of category biomarkers. Cholesterol indicates that some sort of animal product was in the residue, from terrestrial meat, fish, shellfish, or a mixture of these. Tiny amounts of cholesterol are present in some plants (Bianchi et al. 1984), but in such small amounts that cholesterol in measurable quantities in a residue can be confidently used as an indicator of animal products. Plant sterols, such as sitosterol, campesterol, and stigmasterol, indicate that plants were processed in the vessel. Ergosterol indicates some sort of fungal presence, which can come either from fungal contamination of the residue itself or processing a fungal resource in the vessel. There are also many rare sterols, such as occelasterol (algae or shellfish) or avenasterol (plants), that can be used to indicate the classes of items processed in a pot.

There are also *nonsterol category biomarkers*. Terpenoids, discussed extensively in chapter 5, are compounds commonly found in plants. From the plants' point of view, they are usually used to fight insects or infection or to attract pollinators. From a human point of view, they provide the taste and scent of most herbs and spices and the strong aroma of tree sap and pine resin. The lighter terpenoids generally provide most of the scent of, for example, mint, but these lighter compounds do not preserve in most archaeological residues. Only the heavier, more stable terpenoids regularly survive in archaeological residues, in the configurations called diterpenoids and triterpenoids. Pimarane and abietane diterpenoids are indicators of conifer resin usually from pine resin but sometimes from other conifers such as redcedar, cypress, or true cedars. Other types of diterpenoids are less specific but derive from a range of plants. Triterpenoids are found in nonconifer resinous plants. At a minimum, their presence in a residue tells us that plants were processed in the vessel, probably some plant with a strong taste or scent.

Resource-specific biomarkers give the most information about pot use. These are compounds unique to a particular resource, or to a particular resource in a geographic region. For example, some triterpenoids are largely resource specific, such as betulin, distinctive to birchwood. Its presence usually indicates birch-bark flavoring, or the use of birch sap or tar.

A number of *alkaloid biomarkers* will be more extensively discussed in chapter 4. Most analyzed among these resource-specific alkaloid biomarkers are caffeine and nicotine. Nicotine is primarily present in members of genus *Nicotiana* and is usually used as a biomarker for tobacco use. Some non-*Nicotiana* plant species contain nicotine, and in some regions, the presence of nicotine might derive from one of these species (US Department of Agriculture, Agricultural Research Service 1992–2016). Depending on the ethnobotany of the site in question, however, nicotine can usually be used to indicate the presence of tobacco.

Although caffeine is found in several different botanical sources around the world, these sources are generally geographically limited. So caffeine in a residue from China almost certainly came from tea (*Camellia sinensis*), whereas caffeine in the southeastern United States is likely to be from yaupon holly (*Ilex vomitoria*) in the beverage known as Black Drink. Caffeine in Mesoamerica is probably from cocoa, in Brazil from yerba mate (*Ilex paraguariensis*) or guarana, and so on. The difficulty lies in areas that are outside the native habitat of the caffeine sources. Caffeine found in the midwestern United States could presumably derive from either yaupon holly, imported from the coastal regions of the Southeast, or from cocoa, imported from the southwestern United States or Mesoamerica. In this case, the residue analyst looks for additional compounds that might help indicate the source of the caffeine and the analyst will also need to have a conversation with the archaeologist about the most likely source of the caffeine.

Process biomarkers are compounds that form during or after vessel use and that indicate pottery use or postuse heating patterns. At temperatures above 300°C but below 350°C, fatty acids tend to undergo pyrolytic reactions (heat- or fire-induced). Polyunsaturated fatty acids, such as those often found in fish and shellfish, will undergo a sort of fusing of the unsaturated double bonds in the straight chain into an aromatic ring. These pyrolytically formed fatty acids are called ω,o alkylphenyl fatty acids (APHs), and they are extremely stable in the depositional environment (Hansel et al. 2004). This is especially useful because the polyunsaturated fatty acids that form the APHs are prone to oxidation and do not last well in buried residues. The discovery of APHs in a residue indicates both that polyunsaturated fatty acids were present in resources processed in a vessel and that the vessel was heated above 300°C at some point. Since boiling will generally not heat a vessel much above 100°C, the vessel was either accidentally heated/burned, or the vessel was used for some cooking method such as dry cooking or frying, both of which will heat a vessel to the necessary range. Although these compounds may form from plant-derived polyunsaturated fatty acids, they more often seem to form from compounds present in fish or shellfish and are one of the indicators for fish or shellfish processing in a pot. These indicators are discussed at greater length in chapter 7.

Similarly, saturated straight-chain fatty acids will react pyrolytically at about the same temperature, turning into ketones (Evershed et al. 1995). Preservation of the compounds is less important in this case, as saturated fatty acids preserve well in residues. The presence of the ketones, however, indicates that the vessel was heated above 300°C. Depending on vessel shape and the archaeological interpretation, these compounds have been used to interpret a vessel as a lamp (Evershed et al. 1995), or as a griddle, or for frying (Beaman et al. 2016). Although some ketones are naturally present in epicuticular waxes from plants (Evershed et al. 1991), the series of ketones produced by pyrolysis is usually different from that produced by plant waxes (Evershed et al. 1995).

Having described these different types of biomarkers, the unfortunate fact is that biomarkers do not survive in the archaeological record in many residues. Category biomarkers such as sterols are the most common of these compounds to absorb within residues and survive in the depositional environment. In the absence of a biomarker, it is impossible to give a categorical answer as to what was processed in a pot. A good guideline to remember is that resources that contain unique chemicals will often have other unique characteristics—taste, smell, medicinal properties, and such. If a resource tastes and smells bland, it is unlikely to have a unique biomarker other than sterols. This includes most grains and many meats and green vegetables.

Fatty Acid Interpretation

Biomarkers are exciting and definitive, but they are also rare. Most residues are composed of the common compounds described in Table 1.1: fatty acids, alkanols, and alkanes. Of these, fatty acids are the most common. Since these compounds are present in nearly every resource, as mentioned, their interpretation is complicated. Modern resources can be distinguished by the ratios of different fatty acids present in their oil (Malainey et al. 1999a). Olive oil can be easily distinguished from lard. If a residue derives from one resource, such as an olive oil amphora, then the lipids will break down predictably, and fatty acid identification can be used (Malainey et al. 1999b).

Fatty acids must be used cautiously, however, because of the breakdown of shorter and more unsaturated fatty acids described in chapter 1. Fatty acids from fish, shellfish, and plants tend to have more unsaturated fatty acids—that means they have more double bonds in them. These double bonds tend to oxidize preferentially (when fish oil goes rancid, this is essentially the chemical process taking place), so that over time, fatty acids from a fish tend to look more like those from a plant, and then more like those from meat. As a result, oils even from pure resources will not look much like the modern analog, as shown in Figure 1.3.

When working on pottery used to process many different resources, the fatty acids extracted from the residue will come from many different resources at once. In that case, it is not possible to source the residue to one resource based only on fatty acids. More than one resource was present, and the fatty acid mixture can mimic the composition of a resource that was not present in the pot. When in doubt, it's probably best to assume that the vessel was used to process a mixture.

Some general information can be derived from fatty acid compositions, however. They act as an indicator of whether a residue is derived primarily from plants, fish or shellfish, or primarily from terrestrial meat. The composition is best thought of as a sort of average of the fatty acids from resources processed in the pot, with the most abundant contents of the pot dominating the fatty acid composition so that the analyst can say that the vessel was used to contain primarily plants or primarily animal-based products. If the fatty acids contain more $C_{16:0}$ than $C_{18:0}$, and unsaturated fatty acids make up perhaps a quarter of the size of their analog saturated fatty acids, the residue can be described as deriving primarily from plant or fish/shellfish products. If the fatty acids contain more $C_{18:0}$ than $C_{16:0}$, and unsaturated fatty acids are less than a quarter of the size of their analog saturates, then the residue can probably be described as deriving primarily from animal-based products. In a few cases, fatty acids might be so highly unsaturated that they may be cautiously used to indicate fish or shellfish, as discussed in chapter 7.

Case Study 1: Interpreting Your First Residues

The preceding portion of this chapter gives a good overview of how to interpret a residue. It is a lot of information and may seem overwhelming! We will now look at one residue, so that it's easier to see how this information is applied. We will start with a single colonoware sherd excavated from the Brunswick Town/Fort Anderson site in North Carolina by Tom Beaman and Vincent Melomo as part of their Peace College Archaeological Field School. This sherd was extracted with the chloroform/methanol method, producing three separate fractions for analysis, shown in Figures 2.9, 2.10, and 2.11. From the total lipid extract (TLE), shown in Figure 2.9, we can tell that the acyl lipids in this residue have not yet extensively hydrolyzed, with abundant

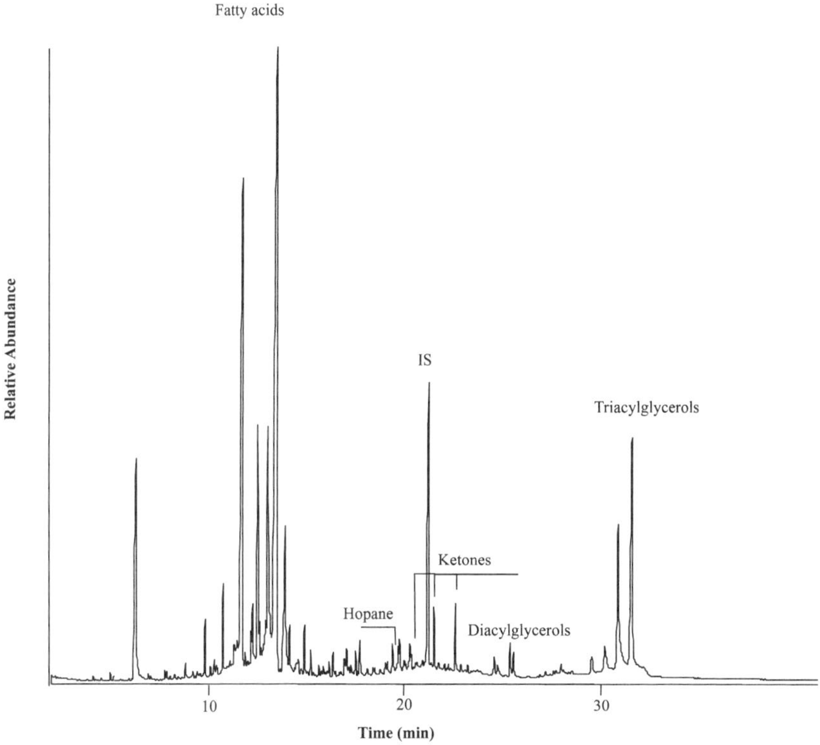

Figure 2.9. Total lipid extract (TLE) of a residue extracted from a colonoware sherd excavated at Brunswick Town/Fort Anderson State Park by Tom Beaman and Vinnie Melomo as part of the Peace College field school. Abundant triacyl glycerols (TAGs) suggest an intact residue. IS = Internal standard. (Eleanora A. Reber)

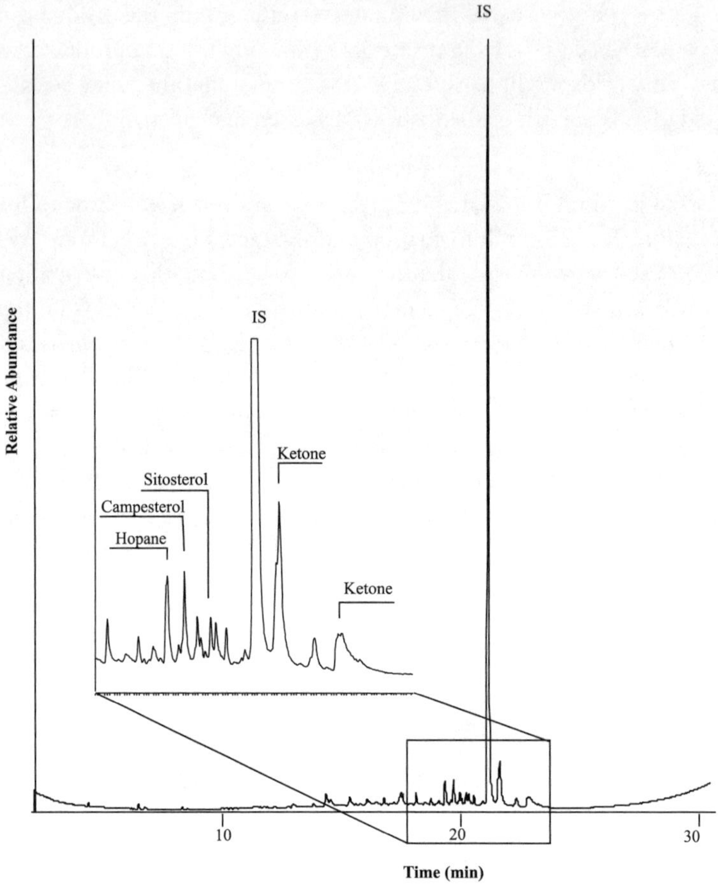

Figure 2.10. Neutral fraction of the residue whose TLE is shown in Figure 2.9. The inset gives details of especially diagnostic compounds identified in the residue. Unlabeled compounds are mostly alkanes. Internal standard = IS. (Eleanora A. Reber)

TAGs as well as fatty acids. This is probably because the residue is fairly recent. From almost the first glance, it is possible to predict that the fatty acids will be not be completely oxidized, leading to preservation of unsaturated fatty acids. The TLE for this residue also contains three ketones that are almost certainly process biomarkers derived from pyrolytic reaction of $C_{16:0}$ and $C_{18:0}$ fatty acids. We know therefore that this sherd was subjected to temperatures between 300°C and 350°C at some point.

In addition, the neutral fraction (shown in Figure 2.10) shows some category biomarkers, including sitosterol and campesterol, both of which are plant sterols, and hopane, a compound often found in petroleum oil, although it may also be found in soil lipids and bacteria.

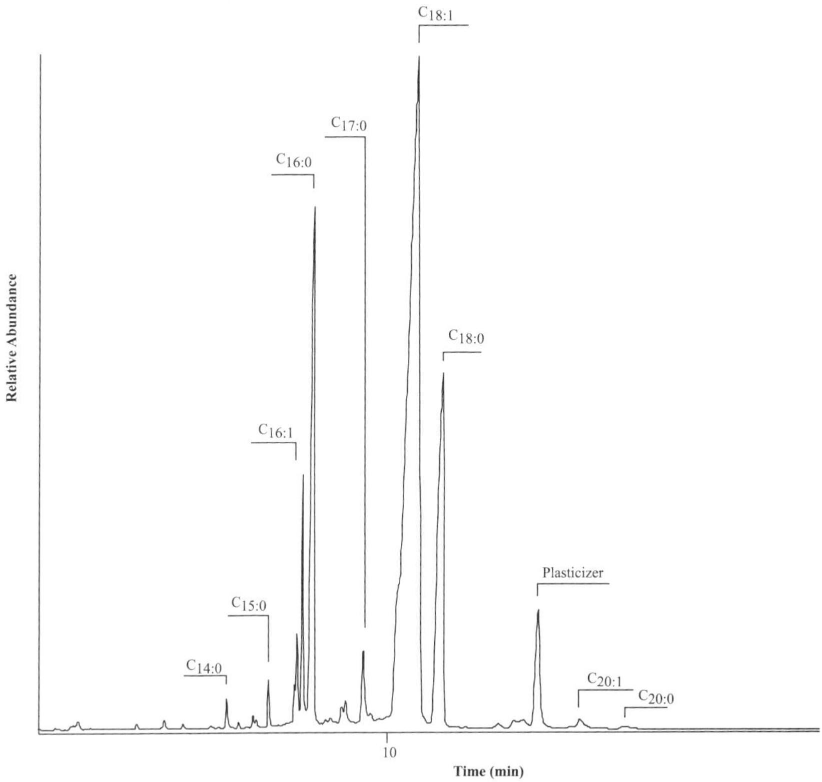

Figure 2.11. Fatty acid fraction of the residue extracted from a colonoware sherd whose TLE and neutral fractions are shown in Figures 2.9 and 2.10. Note that $C_{18:1}$ was the most abundant compound in the fraction and that $C_{16:0}$ was more abundant than $C_{18:0}$. This probably indicates a residue derived primarily from plants, fish, or shellfish. (Eleanora A. Reber)

The fatty acid fraction of this residue, shown in Figure 2.11, can be interpreted as coming from a vessel used to process primarily resources derived from plants or perhaps fish and/or shellfish. The fatty acids (well preserved in the acyl lipids) are dominated by the unsaturated fatty acid $C_{18:1}$, also called oleic acid, while the relatively rare $C_{16:1}$ was also present, as was $C_{20:1}$. Interestingly, a plasticizer, diethylhexyl adipate (DEHA), appeared in the fatty acid fraction that was not present in the TLE or neutral fractions. Presumably, the plasticizer was in polymer form in the TLE, and only became detectable when separated into smaller molecular pieces during the extraction procedure.

The overall interpretation of this residue is that the vessel was used to process primarily plant, fish, or shellfish resources, and it was used in some form of high-temperature cooking, such as frying (due to the ketones). This makes sense given the evidence on the diet of enslaved people in the Brunswick Town/Fort Anderson area, which is known to focus on riverine and gathered resources. Interestingly, local Gullah cooking in the present day often involves fried fish and seafood. The residue also contained hopane, which may derive from some petroleum oil resource or bacteria. This resource might have come from soil pollution; however, we analyzed a comparative soil sample with this sherd, and found no hopane compounds. We could not definitively identify the presence of fish or shellfish in the residue because the residue does not fulfill any of the potential indicators for fish or shellfish in a residue, described in more detail in chapter 7.

A Note on Empty Vessels

One interesting question in residue analysis is how to determine when there are no residues in the pot. The absence of a residue tells the excavator that the pot was unused, or used to contain resources that don't produce residues, such as water or dry contents. During extraction, as mentioned, an internal standard is added that permits the measurement of how much residue is present. This standard is noted as IS in Figures 2.9 and 2.10. In most cases, lipid residues in amounts smaller than 5 μg lipid/g pottery are viewed as too small to interpret, and probably as empty vessels (Evershed 2008b). Depending on the instrumentation and quantification used, smaller amounts may sometimes be interpreted, but it must be done cautiously (Reber et al. 2019, 2020).

USES OF RESIDUE ANALYSIS

In general, residue analysis gives a good idea of the overall resources, and sometimes of specific resources, processed in a pot. However, the analysis cannot identify species of resources that contributed to a residue under most circumstances. Fish and/or shellfish can sometimes be identified, as discussed in more detail in chapter 7, but the individual species of fish or shellfish cannot be determined. Animal resources, such as meat, can be identified, but not usually to species, although isotopic techniques, discussed in chapter 6, can identify ruminant vs. nonruminant animal fats. Resource-specific biomarkers permit identification of specific species in some cases, but these are rare. In most cases, staple grains or starchy foods cannot be identified to species—most exceptions are cases where the resources are isotopically unique, as we'll discuss in chapter 6. Conifer resins can be identified clearly and reliably, but determining the source of the conifer is not usually possible using only residue analysis, as discussed in chapter 5. Sometimes, in conversation with the excavator and a paleoethnobotanist, the most likely source of these resins can be determined. But this information derives from a good knowledge of the environment rather than the residue analysis alone. In short, residue analysis can only rarely identify specific resources processed in a pot. It is a useful tool for an archaeologist but not a magic wand.

3

RESIDUE CONTAMINATION

How to Detect and Avoid It

THE PREVIOUS CHAPTERS largely avoided a key topic in residue analysis: modern contribution to a residue during the time between its excavation and extraction in the residue lab. The issue of contamination is crucial in residue analysis and deserves its own chapter. Because of the sensitivity of GC/MS and HPLC/MS analysis, surprisingly large amounts of modern chemicals can enter a residue by simply storing the sherd in plastic bags, as seen to some extent in Figure 2.11.

Residues, in fact, don't stop forming when they're deposited or when they're excavated. Soil chemistry may affect residues during burial, and archaeological and curatorial activities add to them during processing and storage. These can have a variety of effects on residue interpretation, some quite surprising. Think of these types of contamination as similar to modern artifacts or disturbances on archaeological sites. Usually intrusive artifacts can easily be removed from the top levels of an excavation. Residues, however, do not have stratigraphy, and contamination may sometimes be a serious problem. This chapter discusses types of contamination, how it is sometimes possible to interpret archaeological residues that contain it, and when the residue is so contaminated that it cannot be interpreted.

DEPOSITIONAL CONTAMINATION

In most cases, there is not much contribution to a residue during the time when potsherds are buried—the vast majority of contamination takes place during excavation and laboratory processing. Under normal circumstances, any compound water soluble enough to wash into a residue during burial is also water soluble enough to wash out again. There are rare exceptions to this rule, however. Some stable compounds are found in the soil that are sufficiently water soluble to wash into pottery but insoluble enough to remain there. Often these are lipids such as hopanes from oil or bacteria, and a soil test near the sherd will permit the analyst to determine whether the compounds derived from the soil or the archaeological residue. In rare cases, however, depositional contamination can be overwhelming, as will be discussed. Such cases will usually be obvious to the excavator.

Case Study: The Deepwater Horizon Oil Spill and Its Effect on Pottery Residues

The Deepwater Horizon oil well blowout, technically referred to as the MC252 spill (from Mississippi Canyon Block 252), discharged 210 million gallons of crude oil and 250,000 metric tons of natural gas over a time span of eighty-seven days into the Gulf of Mexico in 2010 (Joye 2015). Due to the catastrophic nature of the spill, dispersants, predominately Corexit 9500 and Corexit 9527, were extensively applied both at the ocean surface on visible slicks and at the wellhead in hopes of dispersing the oil before it reached the ocean surface. The result was fewer visible oil slicks than usual in an oil spill. The dispersants essentially acted like a detergent, breaking the oil particles into small droplets that then traveled through the ecosystem. Some of these droplets accumulated on the shoreline, along river deltas, estuaries, and beaches, where many coastal archaeological sites are located.

Mark Rees and Sam Huey, both of the University of Louisiana at Lafayette, and Scott Sorset of the Bureau of Ocean Energy Management submitted seventeen pottery residues from eight archaeological sites along the Louisiana Gulf Coast. The goal of the project was to see how destructive the oil spill and dispersant release was to pottery residue analysis. To test this, fifteen samples were from sites with visible evidence for oil contamination, while two were from sites without such visible evidence. The answer to the research question is that the MC252 oil spill was very destructive to pottery residue analysis, because of the chemical composition of weathered crude oil and because of the effects of the dispersant.

The most obvious type of contamination was crude oil. Figure 3.1 shows a residue extracted from an eroded Baytown Plain potsherd that is overwhelmed by crude petroleum oil. Crude oil is made of thousands of different complex hydrocarbons. Especially when weathered, it cannot be separated by a standard GC, and the result is a large hump called the Unresolved Complex Mixture (UCM). In this case, we sourced the oil to the MC252 using ratios of diagnostic compounds (Meyer et al. 2018). The sampled potsherd is therefore best interpreted as having been contaminated by oil during deposition, probably from the MC252 spill, rather than being used to process oil

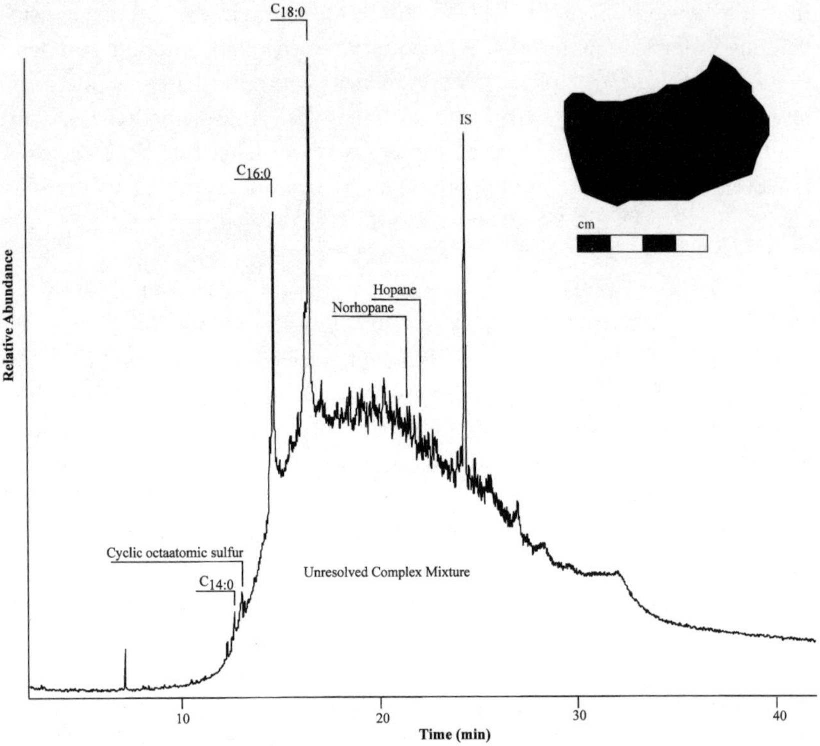

Figure 3.1. TLE of a residue extracted from a pottery sherd from 16SB178, the Southern Comfort site, showing the Unresolved Complex Mixture typical of crude oil. This residue is composed almost completely of crude oil. The eroded Baytown Plain sherd, shown in silhouette, was found on the surface of a beach known to have been affected by the oil slick. (Eleanora A. Reber)

during its use. Since the sherd was found on the surface of a coastal area known to have been affected by the oil slick, this overwhelming oil signature is only to be expected.

It is unsurprising that crude oil contaminated some pottery residues, given the extent of the oil spill and the water insoluble nature of crude oil. More surprising was the effect of the Corexit dispersants. These dispersants were composed of a mixture of compounds, including a proprietary compound called dioctyl sodium sulfosuccinate (DOSS), often used as a biomarker for the dispersants (Kujawinski et al. 2011; Seidel et al. 2016). This compound, however, breaks into smaller, distinctive compounds in the environment that were apparently too water soluble to stay in pottery residues.

When analyzing these residues, however, I noticed that six of the seventeen residues contained an unusual unknown class eventually identified as polyoxyethylene (POE) fatty acid esters, shown in Figure 3.2. Further research suggested that these compounds had been formed by the breakdown of one of the ingredients in Corexit 9500, Sorbitan 80. This compound is known to break down into POE esters (Kishore et al. 2011). Unfortunately, POE esters then break down further into fatty acids. As a result, the fatty acid abundances in an archaeological residue could not be used to interpret vessel contents in any residues showing evidence for this type of dispersant contamination because at least some of the fatty acids likely derived from dispersants. This is particularly impressive because the sherd containing the residue shown in Figure 3.2 was found approximately 30–40 cm below the surface, suggesting that the dispersants migrated through the soil.

Because of the nature of crude oil composition and the composition of the Corexit 9500 dispersant, we realized that the oil spill and its mitigation affected archaeological residues in two different ways. Crude oil includes alkanes and alkanols as well as a wide range of other, complex hydrocarbons. Crude oil contamination was presumably present in any residue showing the hopane and sterane biomarkers for crude oil in this area. Alkanes and alkanols in those residues, therefore, were likely derived at least in part from contamination rather than from human activity in the past. In addition, the Corexit 9500 broke down in sunlight to fatty acids in such a way that a residue showing the POE fatty acid ester biomarkers for Corexit 9500 contamination contained an unacceptably contaminated fatty acid fraction. Any residue showing both types of contamination—and two of the seventeen samples did—was not easily interpretable using any of the standard methods except possibly using sterols to identify category biomarkers.

The good news is that seven of the seventeen samples showed no evidence for any type of contamination, despite the widespread nature of the oil spill and dispersant use. The bad news is that we were not capable of determining a biomarker for the other type of dispersant commonly used in the oil spill: Corexit 9527. This dispersant would also produce fatty acids during breakdown, but without the telltale POE fatty acid esters. One could hope that the two dispersants were used so interchangeably and extensively that where one was found, the other one probably was as well. There is no way to determine this definitively, however, at the present time.

Most archaeological sites will not have major oil spills or dispersant releases nearby, and depositional contamination will not be a major problem. It's worth thinking of the environmental conditions around an archaeological site, though. Large-scale releases of stable, non-water-soluble chemicals into the soil can produce depositional contamination and affect archaeological

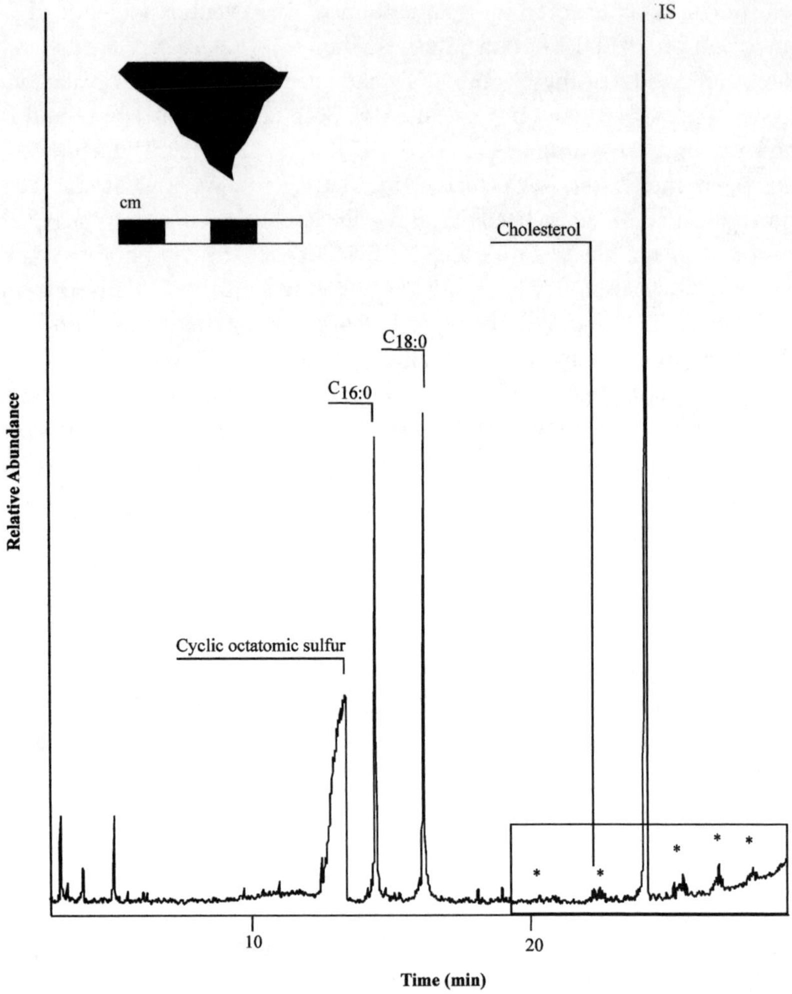

Figure 3.2. GC of the TLE of residue from a grog-tempered sherd excavated from 16LF293, the Redfish Slough site, showing POE fatty acid esters at the asterisked peaks. Although there do not appear to be many of these esters, they break down into fatty acid components. Some of the fatty acids in the residue likely derived from the dispersants rather than archaeological activities. Sherd is shown in silhouette. (Eleanora A. Reber)

residues. We have identified DDT and other organochloride pesticides in residues that clearly derive from depositional contamination, for example. When unsure about depositional contamination, it is often a good idea to take a soil sample from as near as possible to the excavation areas of at least some of the residue samples. This will allow an easy comparison between soil lipids and any unusual lipids found in the residues that might derive from the soil.

DURING AND AFTER EXCAVATION: POSTDEPOSITIONAL CONTAMINATION

Excavation and the ensuing laboratory processing of potsherds can introduce contamination to residues. On nearly all excavations, the excavators will (and should) be wearing sunscreen and insect repellent. These modern substances are engineered to absorb within human skin and therefore often have a glycerol substrate of some sort to encourage absorption. These lipids will overlap with a variety of archaeologically derived acyl lipids in residues and will blend in with them during analysis. This can cause real problems with interpretations. Fortunately, both insect repellent and sunscreen usually also have modern compounds that were not present in the ancient world. When compounds indicative of insect repellent and sunscreen appear in residues, the analyst will need to look closely (and perhaps regretfully) at the fatty acid composition of the residues.

A Note on Sunscreen and Insect Repellent

The active ingredients in sunscreen tends to vary geographically. Sunscreens sold in the United States often use avobenzone, homosalate, oxybenzone, and several other compounds. Sunscreens sold in Europe, the Middle East, and other parts of the world often have different chemical formulations. If you're bringing sherds excavated in Europe to a lab in the United States, your residue analyst will have to keep an eye out for sunscreen components that may be unfamiliar to them. If you are excavating in the United States, and one of your techs is using imported sunscreen, you might want to mention this to your residue analyst. Natural insect repellents that include essential oils but no DEET could potentially cause serious false positives for the residue analyst. If you know people on your crew were using them, it's worth mentioning it to your residue analyst. Fortunately, the sorts of light compounds found in essential oils generally wash out of archaeological residues. Their presence in a residue gives the analyst good cause for suspicion that contamination took place during the digging process.

Furthermore, in many cases sherds are stored in plastic bags at some point in the excavation or curation process. Plasticizers absorb easily into pottery and will often appear in the analyzed residue no matter how careful the analyst is to remove contamination prior to sampling. The plasticizers simply absorb too well within ceramics. Fortunately, plasticizers are polymers and, by the standards of residue analysis, simple and easy to discount. They are usually synthesized by humans in laboratories and didn't exist in preindustrial times. Plasticizers do not usually include acyl lipids, sterols, or other compounds that are interpretively important in residues. As such, they can be discounted from analysis. The primary danger is that a very abundant plasticizer peak may overwhelm smaller amounts of interpretively important compounds. This is called coeluting, and it makes measurement, and sometimes interpretation, of the coeluted archaeological compound impossible.

Ideally, the excavators should select sherds for residue sampling before laboratory processing has taken place, when the sherd has never been stored in plastic. In some cases this doesn't happen, and the archaeologist is interested in sampling sherds that have already been washed in the laboratory, stored in plastic, and perhaps even labeled. This complicates residue analysis but doesn't overwhelm it. In most cases, the labeling compounds will be largely removed by the cleaning stage of absorbed residue extraction described in chapter 2, when the exterior of the sherd is removed to minimize contamination. Some compounds will absorb more deeply into the sherd, however, and will not be removed by this stage. They are also usually distinctive to modern plasticizers and other resources and may be discounted from archaeological interpretation. Still, the more modern compounds absorb within the sherd, the more difficult interpretation becomes. An extreme version of this problem may occur with legacy museum collections, discussed later.

The following steps will help the analyst determine how contaminated the residue might have been during the excavation and storage process. An archaeologist reading a report on residues should keep a sharp eye out for the amount and types of modern contaminants. Be sure to match the information in the report with your knowledge of excavation procedures. Were any unusual chemicals used during excavation that the residue analyst might not have been expecting? For example, at a recent excavation, we used Raid on a vigorous nest of fire ants that colonized our unit. Residue analysis on sherds from this area of the site would have detected some very unusual contamination.

How to Determine Whether a Residue Is Interpretable

A residue whose archaeological signature is overwhelmed by modern contamination cannot be properly interpreted and so is often described as uninterpretable. Here are some steps the analyst will take to determine the interpretability of residues that show biomarkers for contamination.

(1) **Calculating a "percent modern contaminant" measure** is often useful in residues that show abundant contaminating compounds. This may be obtained by totaling the amounts of known modern contaminants (plasticizers, DEET, sunscreen compounds, labeling compounds, and such) and then dividing them by the total noncontaminant compounds in the residue. Although no one to my knowledge has yet come up with definite boundaries of acceptable vs. unacceptable modern contamination, this does give the excavator and analyst a general guideline to work from. If the percent contaminant measure is, for example 50%, then about a third of the compounds in the residue derive from known modern contaminants. This is a lot.

(2) **Looking at the types of contaminants** is helpful as well. If the contaminants are all known modern synthesized plasticizers, the residue is probably still fairly interpretable once these compounds have been discounted. As mentioned, most common plasticizers do not overlap with archaeological compound types. However, if many of the contaminants are from sunscreen or insect repellent, the analyst will need to move to step 3.

(3) **Determining potential modern contamination in the acyl lipids and biomarkers** is the hardest step. As most modern sunscreens and (occasionally) insect repellents are in lotion form, they include many other compounds to help with the absorption of the active ingredients into the skin, as described. These compounds include acyl lipids, fragrances, vitamin E, and botanicals. Most of the fragrances are modern synthetics, so they can be discounted. But acyl lipids are also common in archaeological lipids, and depending on the formulation of the lotion, some sterols may be included. Botanicals will include natural products that could cause interpretive problems such as false positive results for the botanical ingredients.

If a residue includes compounds from sunscreen or insect repellent, or any evidence of modern fragrances or lotion components in step 2, the analyst will need to look carefully at the acyl lipid composition. If they are extremely unsaturated, or if the fatty acids are dominated by $C_{18:0}$ (sometimes used as a component in soaps and lotions), the acyl lipids are probably contaminated by modern products. Depending on the amount of modern contamination (step 1), it is probable that the interpretation of all acyl lipids in the residue is now unsafe, and these compounds must be omitted from interpretation. If the residue also shows light, aromatic botanical biomarkers, of the sort that are usually lost during deposition, they probably derive from modern botanical ingredients in a lotion. It's possible that plant sterols in the residue also resulted from these modern contaminants. In this case, the entire residue is interpretively lost. This is painful! But the alternative is a completely mistaken interpretation. The acyl lipids added to modern sunscreens, insect repellents, and lotions are often highly unsaturated, especially by the standards of archaeological residues, which have undergone oxidation and hydrolysis as discussed in chapters 1 and 2. These modern contaminants will skew the archaeological interpretation strongly toward fish/shellfish/plants, unless the alert residue analyst correctly determines when a residue is uninterpretable due to contamination. Conversely, occasionally modern lotions will use large amounts of a saturated fatty acid. A recent residue analyzed at UNCW contained 400 times more $C_{18:0}$ than $C_{16:0}$. Other biomarkers for modern lotions were present, and I interpreted this unlikely result as due to modern contamination.

Case Study: Detecting and Interpreting Contamination

Examples of more typical residue contamination can be found in residues extracted from pottery excavated at the George Reeves site (11S650) in Illinois, a multicomponent site in the bluffs overlooking the American Bottom, shown in Figure 3.3. The site was intermittently occupied from the Late Archaic through the Mississippian period, with extensive occupations during the eighth to tenth centuries. Parts of the large site were excavated as part of the FAI-270 project in the 1970s and 1980s (McElrath and Finney 1987). A group of entities, including the UNCW Summer Field School and the Powell Archaeological Research Center, excavated another portion of the site in 2002–5. The UNCW Archaeology Lab classes have been assisting in artifact processing and analysis of the site, and sherds are regularly sampled for residue analysis.

Sherds have been sampled under a variety of conditions. Some were selected during excavation in 2004 and 2005. These were generally stored in aluminum foil and/or acid-free tissue paper and weren't washed, labeled, or allowed to touch plastic. Some were selected after the sherds were returned to the lab. At this point, the sherds had been washed, stored in paper bags for some time, and then often stored in plastic bags. And finally, some sherds were selected for sampling after they were fully processed, which included washing, storing in plastic, and labeling with nail polish and india ink. Note that even an excavator who is a residue analyst can select sherds after labeling! It's not ideal, but results can still be obtained if the steps above are used carefully.

Figure 3.3. George Reeves site (11S650) soon after the plow zone was removed. All archaeological stratigraphy was destroyed by plowing, so excavation was by exposed feature. (Eleanora A. Reber)

A good example of this is shown in Figure 3.4. This extract was made from a grit-tempered, cordmarked sherd from the site. The acidic methanol extraction methodology was used, meaning that there is one residue extract from the sherd that shows all fatty acid and neutral compounds but not any of the acyl lipids, as described in chapter 2. There are clearly some plasticizers present in the residue, which is unsurprising as it was sampled after storing in plastic bags and labeling with ink and nail polish. In fact, the plasticizer noted to the left-hand side of the chromatogram (a compound often used in flexible films) and the triphenylphosphate (TPP) probably derive from labeling. The percent contamination is about 21%, which shows a noticeable amount of modern contamination. There are no biomarkers for insect repellent or sunscreen and no light botanical compounds normally typical of hand creams. The ratios of fatty acids are fairly typical for an archaeological residue. The contaminants did not coelute with interpretively important compounds, and nail polish does not normally include fatty acids. Although contaminated, this residue is interpretable.

As it doesn't have any sterol biomarkers, the interpretation rests almost entirely on the fatty acid composition, which includes a small amount of the isoprenoid fatty acid pristanic acid, indicating the possible presence of fish or shellfish lipids. The fatty acids are otherwise not diagnostic of any particular resource and perhaps suggest that it derived primarily from

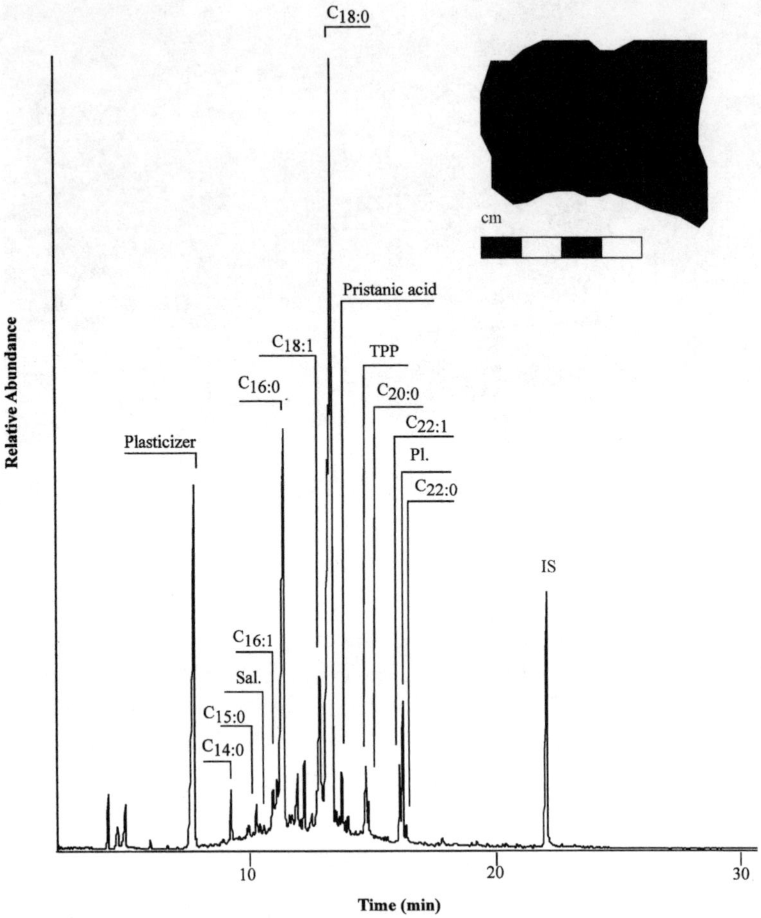

Figure 3.4. GC of a residue extracted using acidic methanol from a sherd from the George Reeves site. The sherd, in silhouette, was a grit-tempered cordmarked body-sherd that had been labeled prior to sampling. TPP = triphenyl phosphate, a common compound in plasticizers and inks. Sal. = salicylate, a common compound found in hand cream and some preservatives. Although salicylate and plasticizer contamination are present, there are no other biomarkers for contamination, and the residue appears to be fairly interpretable despite the labeling. (Eleanora A. Reber)

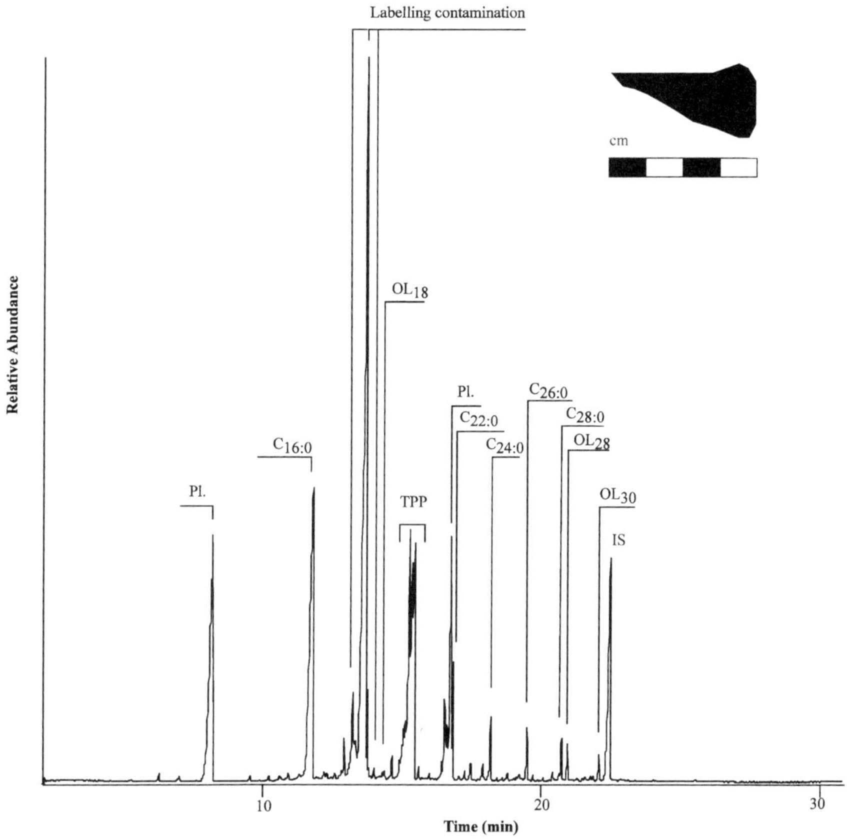

Figure 3.5. GC of an acidic methanol extraction from another labeled grit-tempered, cordmarked sherd from the George Reeves site, shown in silhouette. This residue shows more extensive contamination than the one shown in Figure 3.4, making interpretation difficult. Pl. = plasticizers; TPP = triphenyl phosphate; IS = internal standard. (Eleanora A. Reber)

terrestrial animal products. Although this interpretation is not very specific, it would be impossible to get even that amount of information from the residue if the contamination showed evidence for modern botanical hand cream, for example, which would skew the fatty acid quantification beyond interpretability.

Figure 3.5 shows a more problematic case of contamination. This residue was extracted from another grit-tempered, cordmarked bodysherd from the site, again using the acidic methanol extraction. It also shows plasticizer contamination, including from flexible films and TPP. In this residue, however, the TPP contamination is noticeably more extensive. More importantly, contamination from labeling compounds coelutes with the peak for $C_{18:0}$. This means that the fatty acids can't be quantified easily for this sherd, and an important source of information is lost. This residue tantalizingly contains more neutral compounds than the one shown in Figure 3.4, with the alkanols OL_{28} and OL_{30} suggesting a probable plant presence in the residue. The contamination is so extensive, however, with 58% modern contaminants and coelution over an important fatty acid, that interpretation of this residue is very difficult and probably unwise. The residue is uninterpretable. Sometimes, cleaning the sherd surface before sampling isn't enough to remove most of the modern contamination. This is why residue analysts prefer that sampling take place in the field and directly into tin foil rather than plastic bags, or at least before labeling.

A Note on Legacy Collections

In many cases, pottery may be stored in bags, boxes, or on shelves for some years before the final analysis is complete. At some point in the process, an archaeologist or curator may be interested in residue analysis. A number of factors determine the usefulness of residue analysis on legacy collections: how the sherds were stored, how many curatorial procedures they've undergone through the years, how they were labeled, and so on. Sherds stored in open storage shelves are likely to have undergone complex contamination from the air, as described in chapter 4. Legacy collections from museums where sherds were heavily processed will also be prone to more contamination. When analyzing a series of prehistoric sherds from the McClung Museum at the University of Tennessee at Knoxville, residue analysis commonly found ricinoleic acid, a biomarker for castor oil, in residues. It was unclear where the lipid came from, but as the castor bean is not native to North America, and the sherds had been stored and curated for many years, it seems possible that castor oil was at one point part of the curatorial process at this museum (Reber 2017). The best legacy collections for residue analysis have been stored in paper bags or tin foil with minimal processing or labeling. The second-best collections have been stored in plastic, but different plastic bags break down at different rates. Some bags appear to be in good condition but are actually breaking down and contaminating absorbed and visible residues. If the plastic bags appear to be yellowing and getting sticky, the sherds in them are inappropriate for residue analysis. Residue analysis on legacy collections is worthwhile if the pottery has been stored in closed storage, or if the potential archaeological information from the analysis is important enough to outweigh the possible contamination issues. If samples have been stored on open shelving, it is essential to take swabs of the shelf surfaces to determine the possible extent of air contamination. You'll want to consult your residue analyst on the best methodology for this process.

LABORATORY CONTAMINATION

The excavation process isn't the only time when modern compounds can get into an archaeological residue. Problems in laboratory processing can also arise if the analyst isn't very careful. Lipids are ubiquitous in the human fingerprint, in a single hair from the analyst's head, in poorly cleaned glassware, or in dust in the fume hood. To maintain their sanity, residue analysts have come up with several methods of minimizing this type of contamination and identifying it when it takes place.

All reusable glassware used in residue analysis is first cleaned in alkaline soap and water, rinsed thoroughly (10–15 times) in very clean polished or double-distilled water (tap water has a range of lipids and contaminating compounds in it, especially in Wilmington, North Carolina). It is then heated in a furnace for at least four hours at 400°C. This temperature and heating level will burn any remaining lipids on the glassware to ash, removing it from the glassware. All glass disposable pipettes and tin foil used in the extraction are also ashed in this way. Any items that can't be heated to 400°C are rinsed three to five times with solvent to remove all possible lipids. Gloves are worn during all stages of the extraction procedure to avoid contamination from human touches, and long hair should be kept confined and back from the face.

To ensure that contaminants do not creep into the residues in the lab, archaeological samples are run with a solvent or procedural blank. This empty sample goes through the same lab procedures as the archaeological sample. If no contamination takes place, the blank will be completely empty except for the internal standard when analyzed in parallel with the archaeological samples. If the procedural blank is not empty, some form of laboratory contamination has occurred. Usually, the analyst can determine what sort of contamination took place. If it was from laboratory equipment, the equipment will need to be cleaned and solvent-rinsed. Sometimes, a careless analyst may have touched the glassware with an ungloved finger, or a hair may have floated into the fume hood. A contaminated solvent/procedural blank functionally renders all the archaeological samples in its batch uninterpretable, because the entire batch is assumed to have undergone the same laboratory contamination as the blank. In my lab, we run a procedural blank with every seven archaeological samples, unless the samples are especially valuable, in which case we do one blank to every three archaeological samples. In some studies, the loss of seven samples could be catastrophic.

In general, a laboratory experienced with residue analysis won't suffer a contaminated solvent blank very often—perhaps once every four years. But it is important for the analyst to be ready in case this happens, and it's no bad thing for archaeologists to realize how sensitive this type of analysis can be to contamination.

A Note on Touching Pottery

If a single touch from a ungloved finger on glassware can cause potentially fatal laboratory contamination, what about touching the pottery in the field? At first thought, it seems possible that the simple process of screening in the field could produce enough contamination to affect residue interpretation. Fortunately, there are several procedures during residue extraction and interpretation that help exclude lipids deriving from modern handling in the field. First, cleaning the exterior of the sherd with a model drill appears to remove most lipids deriving from field handling, just as it removes most lipids from sunscreen and bug spray. I believe that this procedure actually works slightly better on fingerprint lipids than it does on sunscreen, because most sunscreens have been specifically formulated to absorb within surfaces, whereas most skin lipids are present as a sort of barrier between the skin and the environment. This has not been experimentally verified, however.

Residue analysts know that removing the surface of a sherd is generally successful in removing modern contamination because recent fingerprints have their own biomarker. Squalene is a highly unsaturated terpenoid compound that is found in a number of species, although rare in mammalian lipids (Downing 1976:24). In commercial use, it usually derives either from shark's livers or from plant forms. It also makes up about 10–15% of human skin lipid (Downing 1976). The compound oxidizes in the presence of air and ultraviolet radiation; the reason it is believed to be present in human skin is to protect it from oxidizing agents in the environment (Huang et al. 2009). Because of its rapid oxidation, squalene from fingerprints will not last particularly long in a residue and is often used as a biomarker, or at least a strong indicator, for recent human handling. This is how fingerprint touches on glassware are detected in laboratory contamination with such confidence. If squalene is found in a residue, it is considered a marker of a recent human touch, either in the lab or the field, and the fatty acids and any cholesterol in the residue need to be carefully evaluated and probably considered the result of this contamination. Squalene is generally absent from residues extracted from sherds cleaned with the drill bit prior to sampling, even when recently handled.

One potential problem with squalene as a marker for recent handling is that it's not entirely clear how long it survives on the surface of archaeological pottery. Forensic studies on fingerprints suggest that it lasts more than a month in dark environments but often less than a month in light environments (Frick et al. 2016). Food chemistry studies suggest that it lasts at least a year in sealed olive oil jars, with more oxygen causing a more rapid breakdown of the compound (Rastrelli et al. 2002). Anecdotally, squalene from field contexts has been detected at least two years after the last known human contact with the sherd. The exact "shelf-life" of squalene has not been published, probably because this time period is highly dependent on the individual skin lipids of the person who touched the surface and the exact storage conditions for the pottery, particularly the presence of oxygen and light (Frick et al. 2016). In general, it would tend to oxidize more quickly in pottery stored in open contexts. Squalene will almost certainly not be present in residues from pottery that was handled in the field more than ten or fifteen years ago, so direct evidence of contamination from handling during excavation will not be present in older collections.

The final way to determine modern fingerprint contamination depends on the analysis of fatty acids, described earlier when determining the amount and type of contamination present in a residue. Modern skin lipids have not undergone the oxidation process common to archaeological residues and are therefore more unsaturated than archaeological lipids. A residue whose fatty acids are dominated by the unsaturated fatty acids $C_{16:1}$ *and* $C_{18:1}$, *with* $C_{16:1}$ *more abundant than* $C_{18:1}$, *could potentially derive from modern human fingerprint oil (Downing 1976). Distinguishing between lipids from a human fingerprint and those from fish or shellfish lipids can be challenging. This is one reason that multiple indicators are usually required for an identification of fish or shellfish lipids in a residue, as discussed in detail in chapter 7.*

CONCLUSION: DON'T PANIC!

This discussion of contamination may seem daunting, given the wide range of modern chemicals that can get into residues. This is not different, however, from modern artifacts that creep into the top layers of an archaeological site, such as shotgun shells and plastic cutlery. In the absence of stratigraphy in residues, contamination is a more serious problem than on an archaeological site. However, like intrusive artifacts at an archaeological site, obvious modern compounds such as plasticizers, DEET, and oil dispersants can be simply discounted from the archaeological residue. The problems tend to arise with compounds present in modern contaminants that are also present in archaeological resources, such as fatty acids or some botanical compounds. In that case, your analyst is going to need to weigh the sorts of contaminants present and determine how much they will affect the interpretation. As an intelligent collaborator with the data, the archaeologist should keep a careful eye on the overall percentage of contaminants and think about how well the findings make sense within the overall environmental and archaeological findings on the site. Dramatic and unexpected findings on sherds with noticeable degrees of contamination should be used carefully and with as much corroborating evidence as possible; this is also the case for sherds without noticeable degrees of contamination.

Finally, when possible, store known residue samples in tin foil, rather than plastic, and try to select residue samples early in the laboratory process before labeling. If you commonly use chemical washes and rinses in pottery processing, such as acid washes and detergents, these will seriously contaminate any residues present and may render them completely uninterpretable.

4

ALKALOIDS AND OTHER MIND-ALTERING SUBSTANCES

ALTHOUGH BIOMARKERS FOR psychoactive substances are rare in the universe of all residue samples, they are of great interest to archaeologists and residue analysts. Most, but not all, psychoactive substances include alkaloid compounds, which are important biomarkers for a variety of different resources. Many of the alkaloids most commonly detected in residue analysis derive from plants, but members of every kingdom of living organism produce these compounds (Aniszewski 2007). The definition of an alkaloid is broad. The term describes any nitrogen-containing ring-structured compound with physiological effects (Aniszewski 2007). These physiological effects are quite wide-ranging; most hallucinogenic compounds such as psilocybin, mescaline, scopolamine, and atropine are alkaloids, but so are caffeine, nicotine, capsaicin, and other compounds that are only mildly psychoactive. All opioids and cocaine are also alkaloids. To make matters more complicated, alkaloids are so notorious as mind-altering compounds that the term is sometimes used broadly to describe almost any ring-structured compound with psychoactive effects. For example, cannabinoids are sometimes classified as alkaloids despite the absence of nitrogen in these compounds. From a residue standpoint, cannabinoids are phenolics and are briefly discussed in chapter 5.

This chapter describes extraction and analysis procedures for some of the better-known and more common alkaloids of interest in residues. A surprising number of alkaloids have not yet been investigated through residue analysis, however, probably due to lack of communication between archaeologists and residue analysts. If you are interested in a particular alkaloid or psychoactive compound not described in this chapter, you should start talking with a residue analyst. It should be relatively easy to determine the optimal extraction and analysis for that compound, and the results could be interesting.

Although all alkaloids have some physiological effect, there are a range of other psychoactive substances that are not alkaloids. This chapter also discusses the detection of various types of alcoholic beverages because of the importance of these mind-altering substances in archaeological interpretation. Finally, contamination in the field and lab can be a serious problem with some psychoactive biomarkers. This chapter provides an overview of when this type of contamination may be an issue and how to avoid or quantify it.

AN OVERVIEW OF ALKALOIDS

The physiological effects caused by alkaloids often include psychoactive and medicinal properties, and so the presence of alkaloid biomarkers can identify use of mind-altering substances, medicinal practices, and probable ceremonial behaviors in the past. Among the alkaloids that have been identified using residue analysis are nicotine, caffeine, theobromine, theophylline, and atropine. Given the physiological properties of alkaloids, however, the presence of any alkaloid in an archaeological residue is of extreme interest in interpreting the behavior of past peoples.

Unfortunately, the chemical structures included in the alkaloid category are so broad that there is no single way to extract all alkaloids with one methodology. Most alkaloids are alkaline (hence the name), but some are not. Many of the compounds are insoluble in water and are soluble in organic solvents, but others, including the important biomarkers caffeine and theobromine, are soluble in water, and theobromine is only reasonably soluble in hot water, ammonia, and ethanol. Further, as alkaloids are not lipids, the standard lipid extraction procedure will not always work on them. Sometimes it will, but it varies based on the individual compound.

This has a twofold effect on communication between archaeologists and residue analysts. First, unless the archaeologist tells the residue analyst that they are interested in a particular alkaloid compound, the residue analyst will not use an extraction methodology tailored to the recovery of that compound. Second, because alkaloids are so rare in residues overall, the residue analyst might easily forget to mention to the archaeologist that some alkaloids will not be recovered using the standard lipid extraction methodology. If an archaeologist is interested in an alkaloid compound, they should specifically tell the residue analyst, who can then check to see whether that compound is extracted using standard methodologies, or whether a different extraction procedure will need to be used. If the archaeologist is interested in both standard lipid residues and in looking for an alkaloid that requires a special extraction technique, the sherd will probably have to undergo two extraction procedures. In this case, the sample will be divided into two parts, each one of which will be extracted using a different technique, so that all compounds of interest are extracted. This procedure will increase the time and cost of analysis and will require a larger pottery sample for destructive analysis, in the case of absorbed residues, or a larger amount of visible residue, for visible residues.

Because of the broad range of alkaloids, the archaeologist is going to have the best sense of which alkaloids may be of interest in a particular region or pottery type. The residue analyst will probably not know, for example, whether psilocybin mushrooms are native to the area of excavation, if poisonous frogs or other amphibians were commonly used in some way, or whether another alkaloid-containing psychoactive was culturally important to the people in the region. In addition, as the method of introducing such alkaloid psychoactives into the body varies culturally from place to place (smoking, ingestion, etc.), the analyst won't know the likeliest way in which the alkaloids might have been processed. This area of investigation requires close cooperation between the archaeologist and analyst to get good results.

A Note on Choosing Samples for Alkaloids and Psychoactive Substances

As the archaeologist, you are the best judge of the sorts of ceramic items that should be sampled for various alkaloids and psychoactives. That stated, here are some general ideas that might help prime your thinking. Pipes, for examples, are an obvious item to sample for nicotine, but also potentially for other alkaloids. Keep in mind that not all pipes were used to smoke tobacco. At least in North America, indigenous peoples often smoked a mixture of herbs that sometimes included tobacco (Gilmore 1977; Rafferty 2016). Other alkaloids, such as opium, are also commonly smoked.

Pots with a very unusual shape or size are also a good place to start. For example, pots from west Mexico decorated like datura seed pods suggested to Adam King and collaborators (King et al. 2018) the potential presence of the psychoactive plant in residues. Decorative motifs reminiscent of datura flowers suggested the use of the psychoactive plant at Pinwheel Cave in California, which was later confirmed by a series of methods, including residue analysis of chewed quids (Robinson et al. 2020). Mimbres pots with decoration suggestive of possible datura use are shown in Figures 4.1 and 4.2.

Pots in the shape of shell cups, as shown in Figure 4.3, have been used as a potential indicator for the use of Black Drink, a ritually important caffeine-containing tea made in the southeastern United States from yaupon holly (Ilex vomitoria). This is because of the known cultural connection in the Southeast between large shell cups and the serving of the purifying beverage (Milanich 2004). Similarly, much of the early work on

Figure 4.1. Mimbres small vessel with applied protrusions found in Grant County, New Mexico, very similar to a datura seed pod (Berlant and Maurer 2018). (Museum Purchase, 1930. Image © President and Fellows of Harvard College, Peabody Museum of Archaeology and Ethnology, Harvard University 30-5-10/97686)

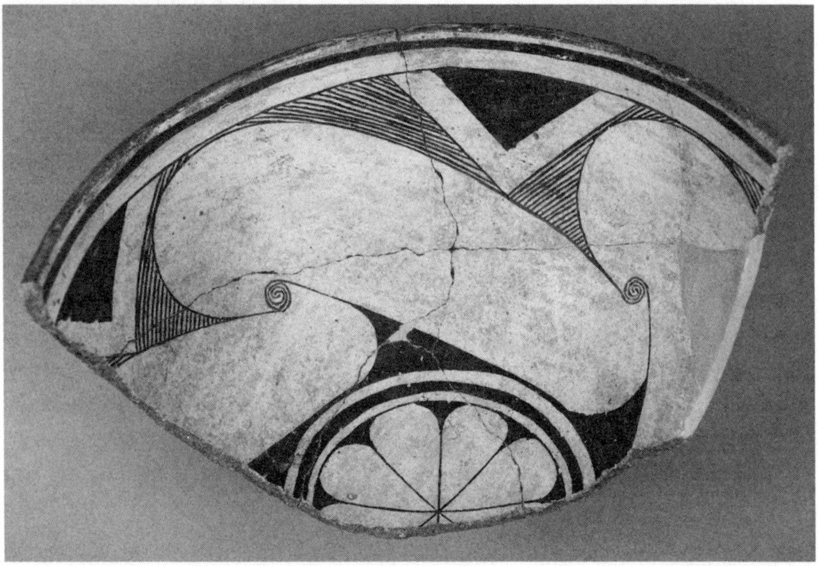

Figure 4.2. Classic Mimbres bowl from Grant County, New Mexico, decorated with swirls reminiscent of a datura flower. This vessel and that in Figure 4.1 demonstrate the sorts of motifs that would suggest sampling for datura use (King et al. 2018; Robinson et al. 2020), although neither vessel has been sampled. (Peabody Museum Expedition, C. B. and H. S. Cosgrove, Directors, 1924–1925. Image © President and Fellows of Harvard College, Peabody Museum of Archaeology and Ethnology, Harvard University 25-11-10/94864)

Figure 4.3. Carson red-on-buff *Busycon* shell effigy vessel from the Parkin site in Arkansas. Although this vessel has not been sampled for yaupon use, this type of shell effigy is considered appropriate for investigation of Black Drink use (Milanich 2004). (Peabody Museum Expedition, E. Curtiss, Director, 1878–1879. Image © President and Fellows of Harvard College, Peabody Museum of Archaeology and Ethnology, Harvard University 79-4-10/20191)

wine detection focused on amphorae, which seemed likely to contain either wine or olive oil (Badler et al. 1990; Michel et al. 1993, and others).

Ethnographic research on behaviors or ceremonies around psychoactive plants or alkaloid-containing foods is a helpful resource when determining which vessel types should be sampled for mind-altering compounds. For example, ethnographic writings on Black Drink suggest that the tea was boiled in large straight-sided cooking pots and then ladled into the shell cups (Milanich 2004). Such large straight-sided pots, if found, are then a good candidate for caffeine analysis. Of course, from an archaeological standpoint, the pots could also have been used to make stew or any other large amount of food. Residue analysis can help answer this question as long as the analyst knows that the question is being asked.

TARGETED VS. UNTARGETED ANALYSIS

Untargeted analysis is the type of residue extraction and analysis already discussed, in which the goal is to detect and identify as many compounds, usually lipids, in the residue as possible. Targeted analysis is a method with extraction and instrumental analysis specially developed to optimize yield and detection of one, or a small number of, compounds of interest. In many cases, the archaeologist and residue analyst are not expecting to find an alkaloid biomarker in their residues. Untargeted analysis is really the only option in that case. In the best case, untargeted analysis allows the detection of compounds that are completely unexpected in the submitted residues. This kind of serendipity is one of the things that makes residue analysis exciting! Untargeted analysis, however, will almost never produce the sensitivity found in targeted analyses. In addition, some alkaloids will not be extracted or detected by one or more of the standard extraction methodologies, and so untargeted analysis will not work.

When the archaeologist is interested in looking for particular compounds, targeted analysis is often a good option. The decision to use targeted methods involves a trade-off, or series of trade-offs, however. Targeted extraction methods often involve functionally sampling a sherd twice, which is time consuming and uses twice as many materials as a single extraction. It will likely involve two separate instrumental runs on two separate instruments. Many labs are set up optimally to use one or another type of chromatography, often GC/MS for untargeted residue analysis. A combination of targeted and untargeted analysis is therefore more time consuming, costly, and destructive than one of these methods alone, and will require access to a wider range of equipment. Applying targeted analysis alone will be cheaper, quicker, and less destructive than using both methods, but the information about overall vessel contents obtained by untargeted analysis will be sacrificed. The analysis will only recover information about the targeted compound or compounds. Since the excavator is most familiar with the importance of the archaeological questions involved, it's important to understand the costs and benefits of combining these approaches or selecting one of them. Below are some different ways of approaching targeted and untargeted analyses to obtain as much information as possible about small quantities of important biomarkers for mind-altering substances. Most of these techniques will also work on any biomarker.

Solubility and Extraction Method

When interested in an alkaloid or biomarker for psychoactives, but attempting to use untargeted analysis, it's important to determine whether the compound is soluble in one of the standard solvent mixtures for residue extraction. If so, it's best to choose whichever solvent mixture dissolves the target compound most effectively. If the compound is soluble in chloroform but not methanol (as is the case for caffeine and theophylline, discussed further on), then targeted and untargeted analysis can be combined to an extent, as long as the chloroform/methanol extraction method is used rather than the acidic or alkaline methanol extraction methods. If this is the case, the best approach is probably to use untargeted analysis as a screening step to determine whether the compound is present in a residue while also obtaining information about the overall residues. If the compound of interest is present, targeted analysis can then be used to obtain better, more sensitive results. If the alkaloid or other biomarker does not dissolve in any of the standard solvent mixtures, then untargeted analysis cannot be used to detect the compound. Targeted extraction and analysis is the only option. The solvent for targeted analysis should be chosen specifically around the solubility of the target compound(s).

One of the best known of these targeted extraction methodologies is the hot water extraction briefly described in chapter 2. It is particularly effective on compounds that are more soluble in hot water than in chloroform/methanol or methanol. As water is not an optimal solvent for most GC/MS systems, it is often paired with HPLC/MS analysis. Targeted extraction methodologies can use almost any solvent or solvent mixture, however, as long as it dissolves the target compound(s).

Any targeted extraction, even if not using water, is often well suited to HPLC/MS analysis, because only a small number of compounds of interest need to be dissolved in the mobile phase. With the A and B solvents in the HPLC specifically chosen to optimize the separation of the target compound(s), the HPLC/MS will usually be able to detect the compounds of interest with greater sensitivity and quantifiability than GC/MS. The standard lipid extract, consisting of many different compounds with different solubilities, is usually best analyzed with the GC/MS, as described earlier.

Extracted Ion Chromatographs

The extracted ion chromatograph (EIC) method is not a targeted form of extraction or sampling but rather a kind of transitional way for the analyst to focus on a compound of interest in either untargeted or targeted analysis. A standard chromatograph plots the abundance of all the molecular weights that the mass spectrometer is measuring. Small amounts of some weights may be overwhelmed by background noise, especially in complex mixtures. However, when a compound has a unique molecular weight fragment in its mass spectrum, the data system can plot a chromatogram of only that weight from the original data. This is called an EIC because only one fragment is extracted from the rest of the data. The other weights will still be measured, and can also be plotted, so that no data is lost. Some of the sensitivity in HPLC/MS analysis of alkaloids comes from EIC plotting the peak of interest. For theobromine, for example, 180 is the most important weight and is extracted to locate a probable peak. The overall mass spectrum of that peak is then examined, and the presence of theobromine is confirmed from the overall spectrum. Without EIC plotting, the location of the actual theobromine peak could be lost behind other, more abundant compounds in the residue mixture.

The same technique can be used for GC/MS. If an analyst is looking for a particular compound, it is common to run an extracted chromatogram of just one or two mass weights of interest. The resulting chromatogram will allow the analyst to find potential peaks of interest, which can then be checked against the overall mass spectrum. EIC, in short, is a way of finding very small peaks in a residue, as long as the analyst knows what they're looking for.

Selected Ion Monitoring

The most dramatic way to increase sensitivity for a targeted compound in any MS analysis is selected ion monitoring (SIM). In standard quadrupole mass spectrometers, the mass spectrometer is in scan mode. That means the instrument is measuring a range of masses, often from 50 to 600 (or so) amu. Quadrupole mass spectrometers can only measure one weight at once. In scanning mode, then, the instrument is essentially running through a range of 550 or so different masses at a measurable but quick speed (perhaps 0.1 second). It is much quicker and more accurate for the mass spectrometer to

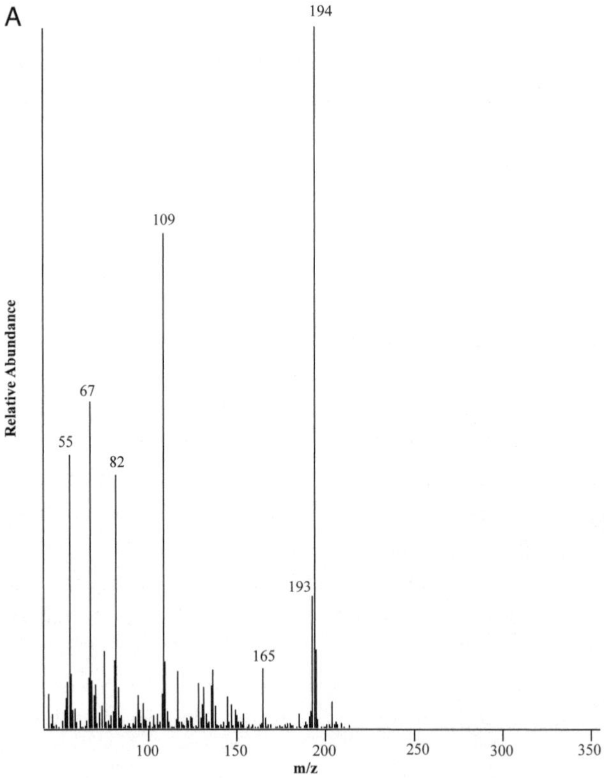

Figure 4.4. A, left, mass spectrum of caffeine in scan mode. (Eleanora A. Reber)

measure only a small number of masses. This setting is SIM mode. For example, if the analyst is interested in caffeine, the mass spectrum of this compound has a large peak at 194, and also strong peaks at 109, 55, 67, and 82, as shown in the top portion of Figure 4.4. If the analyst is only interested in caffeine, a SIM mode focusing on these five weights will be much more sensitive

Figure 4.4. B, right, mass spectrum of $C_{16:0}$ in scan mode. (Eleanora A. Reber)

Figure 4.4. C, below, mass spectra of the same $C_{16:0}$ peak in SIM mode scanning for masses important to caffeine only. Note that although the SIM fatty acid spectrum is not similar to the caffeine spectrum, on a cursory inspection it looks more like caffeine than $C_{16:0}$ because of the nature of SIM mode. (Eleanora A. Reber)

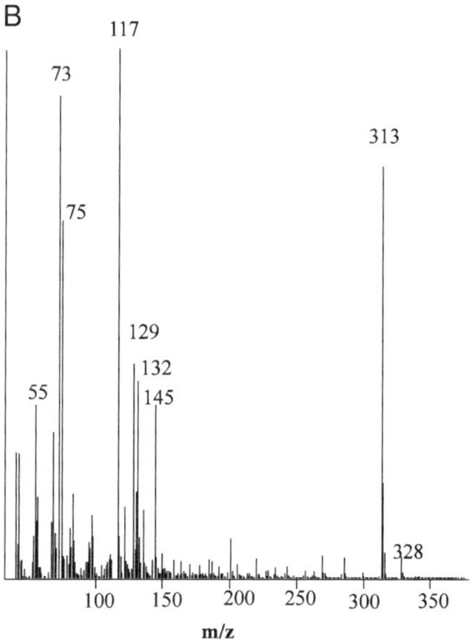

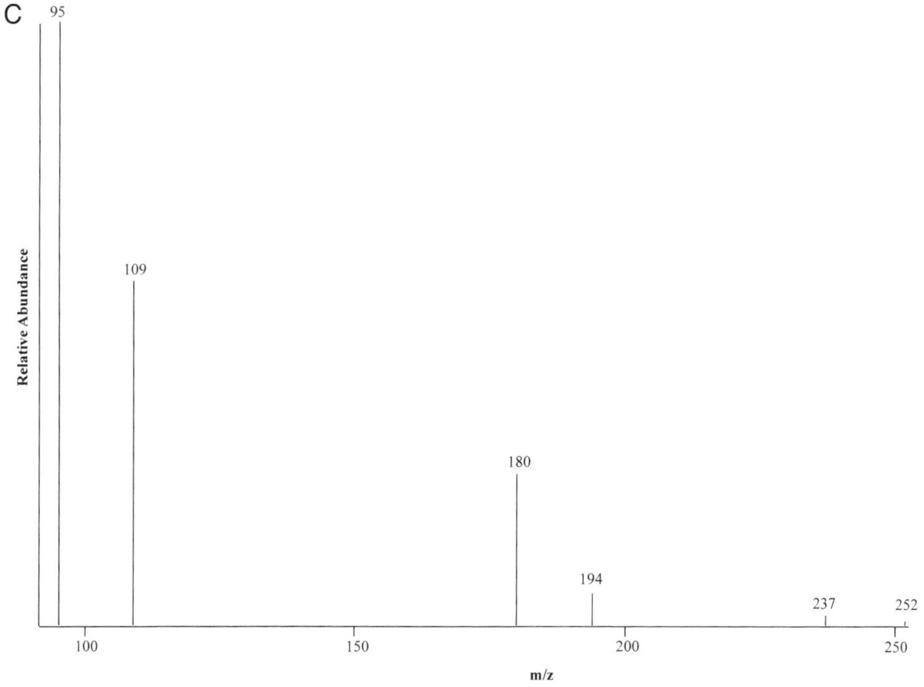

than a scan mode. Assuming a normal scan mode of 550 amu, the 5-scan SIM mode will be about 110 times more sensitive than the scan mode—as long as the instrument is only looking for caffeine.

When the analyst knows the elution time of peaks of interest, the system can be programmed so that the MS is scanning for, for example, caffeine for one time interval, and other compounds at other times. This allows very sensitive analysis of several compounds in a single sample. SIM monitoring does have one weakness. Because it is only looking at the weights of interest, it will never see any other molecular weights in the residue. Every peak detected by the MS in this mode will look at least a bit like the peaks of interest, because the equipment is only set to find those peaks. If this seems confusing, observe Figure 4.4, which shows the scan MS of caffeine, the scan MS of the common fatty acid $C_{16:0}$, and the same peak in SIM mode when scanning for caffeine. This is why studies using SIM mode generally rely on several different methods to confirm the MS identification. The analysts usually run the sample in scan mode as well as in SIM. If the compound is present in large enough amounts to be detectable in scan mode, this confirms the SIM result. The compound may not be present in large enough quantities for this type of confirmation, however. In that case, the confirmation may be via elution time. If the compound looks, for example, like caffeine and elutes at exactly the same time as a known sample of caffeine, it probably is caffeine. In cases where certainty is absolutely necessary, the sample can be confirmed by an external spike. That is, the sample is first run as extracted. If a potential target compound is found in small amounts, the analyst adds a tiny amount of the known compound to the sample. If the known addition coelutes perfectly with the archaeological compound, they are almost certainly the same, and the identification is confirmed. If the two compounds form two slightly different peaks, then they are two different compounds, and the original identification was incorrect. Once the external spike is made, however, the archaeological sample is purposely contaminated.

Scan and SIM modes are unique to quadrupole mass spectrometers, which run through a range of mass weights in turn, as already described. Time-of-flight (TOF) mass spectrometers actually measure all molecular weights at once. This type of instrument, when attached to a GC or HPLC, can provide SIM-type sensitivity with a scan-like mass range. It is probably preferred when looking for small amounts of target compounds but is sometimes unavailable.

SAMPLING FOR CAFFEINE AND OTHER METHYLXANTHINES

Caffeine is found in a range of plants around the world, where it apparently acts as a pesticide. In archaeological residues, it is a geographically based biomarker, as it occurs in various plants worldwide, but the culturally used plants in a particular area are usually unique. For example, coffee and kola originate in Africa, tea in Asia, chocolate in Mesoamerica, yerba mate and guarana in South America, and yaupon holly and dahoon holly in the southeastern United States. Caffeine in a residue from North Carolina likely derives from yaupon holly (*Ilex vomitorius*), while the same substance in a residue from eastern Asia would probably come from tea (*Camellia sinensis*). There is some difficulty in areas located between two potentially importable caffeine sources, discussed in the note below. Caffeine in residues from the midwestern United States could derive from either yaupon holly originating in the southeastern United States or from chocolate originating in the Southwest.

Structurally similar compounds, called methylxanthines, often co-occur with caffeine (Figure 4.5). Theobromine and theophylline are among the most common noncaffeine methylxanthines in archaeological residues. Theobromine is the most abundant methylxanthine in chocolate, with caffeine and theophylline occurring in smaller amounts (de Paula Lima and Farah 2019). Small amounts of theobromine and theophylline also occur in tea, while theobromine occurs with caffeine in kola nut and yaupon holly. As is the case with all compounds used as pesticides by their source plant, the amounts of caffeine and other methylxanthines can vary dramatically within a species. This variation may be based on the age of the leaves or other caffeine-containing part of the plant, the sex of the plant, or the number of problematic insects in the growing area.

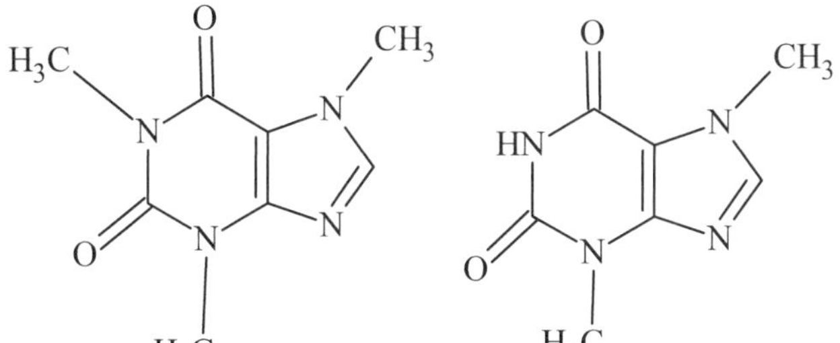

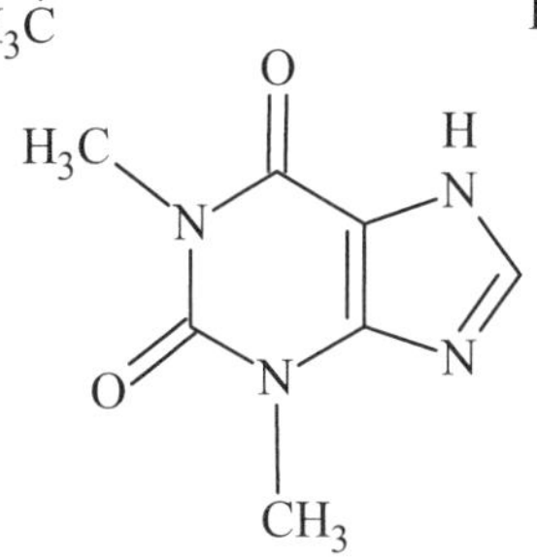

Figure 4.5. Chemical structure of caffeine and other methylxanthines of archaeological interest. Caffeine is in the top left, theobromine in the top right, and theophylline below. Note the presence of ring structures and nitrogen, both typical of alkaloids. (Eleanora A. Reber)

The molecular structures of all the methylxanthines lend themselves to complexing with other compounds. Complexing occurs when weak chemical bonds occur between two molecules. They are important for residue analysis because methylxanthine complexes have different solubilities than the uncomplexed compound. For example, caffeine is highly soluble in chloroform—noticeably more soluble than it is in water. But it is impossible to extract caffeine directly from tea leaves with chloroform because the caffeine has formed complexes with tannins in the tea (Onami and Kanazawa 1996). To remove the caffeine, the tea must either be brewed with water and then extracted, or be chloroform-extracted with an alkaline pH. Hot water extraction works because the caffeine complex is soluble in hot water, and the bonds in the complex loosen in solution. The brewed tea can then be extracted with chloroform to obtain the caffeine. In the direct extraction with alkali solvents, the pH change weakens the complex, and the caffeine can be directly extracted from the leaves. A similar effect happens with caffeine and chlorogenic acid in yaupon holly, as seen in the case study below.

As caffeine is soluble in chloroform, and one standard mixture for extraction of residues is 2:1 chloroform/methanol, caffeine may be detected using untargeted analysis, as long as the analyst uses chloroform/methanol extraction (Reber and Kerr 2012). Since caffeine is sensitive to pH changes and is less soluble in methanol than in chloroform or water, the extraction methodologies using direct acidic and alkaline methanol extractions will extract caffeine less successfully than the chloroform/methanol extraction methodology. The same is the case for theophylline, which is also soluble in chloroform as well as water. If the archaeologist is interested in extracting theobromine as well as caffeine and theophylline, the situation changes. Theobromine is not soluble in chloroform, although it is slightly soluble in methanol. In fact, theobromine is an unusually insoluble compound, being soluble only in hot water, ammonia, and ethanol. Hot water extraction and HPLC/MS analysis is the best method for its detection and analysis.

A Note on Why Theobromine Is Important

Theobromine is a rare compound, present in large amounts primarily in chocolate and in trace amounts in several other caffeine sources already described. Extracting and analyzing it require a targeted extraction method and HPLC/MS analysis, unlike caffeine and theophylline, which can be detected using untargeted analysis. When is it worthwhile to go through this trouble? Primarily, when the archaeologist is trying to distinguish between cacao and another caffeine source. Note that this area of residue analysis is still in development and is often surprisingly complicated, although productive of interesting results.

For example, caffeine and theobromine have been found in residues from Cahokia (Crown et al. 2012). Because this location is out of the native ranges of cacao and yaupon, the methylxanthines could have derived from either source, traded from either southwestern or southeastern North America. In cacao, theobromine is the primary methylxanthine, being present in an 8:1 theobromine:caffeine ratio (Tarka and Cornish 1982). In yaupon, caffeine is the primary methylxanthine, with small amounts of theobromine present in an approximate ratio of 1:6 theobromine:caffeine. In general, residues containing more theobromine than caffeine would be more likely to derive from cacao than yaupon; however, there are two complications.

First, dahoon holly, Ilex cassina, is also native to the southeastern United States and contains both caffeine and theobromine. The compounds are present in much smaller amounts than in yaupon holly, and yaupon was the plant historically associated with Native American caffeinated beverages in the coastal Southeast (Hudson 2004). A study by Edwards and Bennett (2005) concluded that dahoon holly contained caffeine and theobromine, but in insufficient quantities to act as a central nervous system stimulant. This plant is nonetheless a possible source of methylxanthines, and it has a ratio of theobromine:caffeine of approximately 2:1 (Edwards and Bennett 2005).

Second, as caffeine is notably more soluble in cold water than theobromine, it is likely that more caffeine would wash out of a pottery residue during burial than theobromine. Depending on the amount of rainfall and the specific soil chemistry (always difficult to predict), it is possible that the relative amounts of caffeine and theobromine in a residue could reverse during the burial process. No experimental work has yet taken place on this subject. The question is complicated even more when one considers that methylxanthine concentrations and ratios vary dramatically between individual plant sources in the same species, depending on age of leaves, growing habitat, and so on.

The presence of these complicating factors made the identification of cacao vs. yaupon or dahoon holly in the Cahokia residues difficult to accomplish only through methylxanthine extraction. The authors of the study therefore looked for other compounds that are present in holly leaves but not in cacao. Identifying ursolic acid in the residues, the authors argued that this compound suggested that the methylxanthines likely derived from yaupon and/or dahoon holly, both of which contain ursolic acid, rather than cacao, which does not contain it. That said, ursolic acid is present in a wide range of other plant resources in the midwestern United States, including grapes and a variety of leaves often used medicinally (US Department of Agriculture, Agricultural Research Service 1992–2016). As long as the pottery from Cahokia was only used for caffeinated beverages, Black Drink produced from yaupon or dahoon holly seems the most likely interpretation. If the vessels were used for a mixture of resources, the interpretation becomes more complex. However, if the study had tested only for caffeine, instead of caffeine and theobromine, these interesting issues in methylxanthine origin and preservation would not have been uncovered at all.

Case Study: Experimental Production of Yaupon Residues

When I first moved to Wilmington, I was naturally curious about this native source of caffeine endemic to the area. In the years before widespread internet, it took several months before I realized that not only was *Ilex vomitoria* endemic to the Wilmington area, it was also one of the most popular landscaping shrubs in the region. It is suited for sandy soils, drought tolerant, salt tolerant, and evergreen and produces attractive red berries in the winter (Figure 4.6). My students and I immediately set out to make yaupon tea (Black Drink), analyze it, process the tea in experimentally produced pottery, bury sherds from the pottery, extract the absorbed residues from the buried pottery, and then analyze them.

In our first studies, we made tea with a variety of yaupon leaves from several different locations around Wilmington. As expected from the literature,

Figure 4.6. Yaupon holly (*Ilex vomitoria*) with berries. (Eleanora A. Reber)

there was a range of caffeine levels in plants from different environments. We concluded that the highest caffeine levels in yaupon that we could easily harvest were found in a shrub located in my backyard (a very moist, buggy environment with minimal pesticide or gardening intervention) and in a hedge located near the UNCW science building (Dobo Hall). All later studies used these two sources of yaupon, at least until Dobo Hall lost its landscaping during Hurricane Florence.

In a second phase with then-undergraduate Matthew Kerr (Reber and Kerr 2012), we made yaupon tea and boiled it repeatedly in a series of pottery vessels helpfully produced by members of the UNCW Anthropology Club. All vessels processed the same tea, for the same amount of time, over the course of 20 brewing episodes. The pots were then broken, divided into five groups, and placed in different environments. One group was buried in the swampy portion of my problematic backyard, shown in Figure 1.4, while another was buried in the dry portion. A third group was buried in a completely different environment in the Sandhills (thanks to Daniel Autrey of Fayetteville State University), and a final group was placed in a freezer to control for interactions with soil, which we expected to be significant. Sherds were excavated after two months, shown in Figure 4.7, and then analyzed.

Originally, we thought that the caffeine might wash out of the potsherds and be lost relatively early in the burial process. After all, the compound is clearly water soluble and has a half-life in the body of slightly less than six

Figure 4.7. Experimentally produced sherds with yaupon tea residues after Matthew Kerr excavated them from dry, sandy loam soil after two months of deposition. (Eleanora Reber)

hours. We underestimated the tendency of caffeine to complex with other compounds, possibly including clay particles, and also the difference in caffeine solubility between hot and cold water.

Upon extraction, it was clear that the relative amount of caffeine was actually increasing in absorbed pottery residues through time, instead of decreasing. As shown in Figure 4.8, caffeine is present in freeze-dried extracted tea but appears more abundant in an extracted sherd. This is not because the amount of caffeine is actually increasing, but because the other tea compounds either did not absorb into the pot wall in the first place or were lost more quickly than caffeine during burial. The analysis of the freezer sherds demonstrated that many of the compounds in the tea never absorbed within

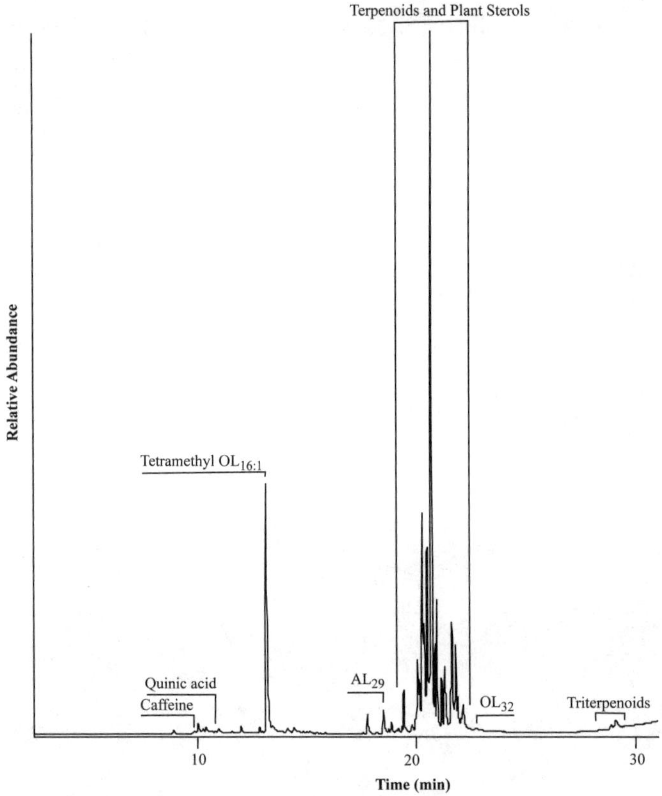

Figure 4.8. Neutral fraction of lipids extracted from yaupon tea. Many of the compounds present in this fraction did not absorb within the sherd, or washed out of the pot during burial, as seen in comparison with Figure 4.9. (Reprinted from Eleanora A. Reber and Matthew T. Kerr, "The Persistence of Caffeine in Experimentally Produced Black Drink Residues." *Journal of Archaeological Science* 39[7]:2312–2319, 2012, with permission from Elsevier)

the pottery at all, while caffeine did (Figures 4.8 and 4.9). Many of those compounds visible in the freezer sherd were either soluble in rainwater and lost during burial or underwent oxidation, hydrolysis, and/or microbial breakdown as discussed in chapter 1.

Further, caffeine was more abundant in the neutral fraction than the total lipid extract. It seems likely that caffeine was complexing with chlorogenic acid also present in the tea, and only some of it was visible in the GC/MS until saponification, which would have separated the complexed compounds, freeing the caffeine. This probably also explains why the caffeine persisted in the sherd longer than we originally expected.

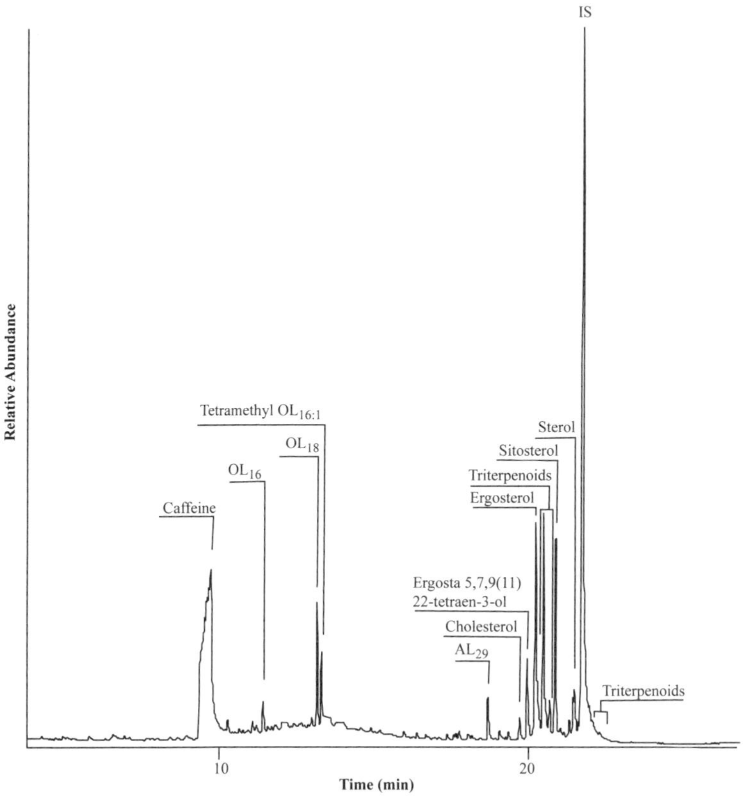

Figure 4.9. Neutral fraction of lipids extracted from a pot used to process yaupon tea and then buried in a sandy noncoastal environment in the North Carolina Sandhills. Note the increase in the relative size of the caffeine peak from that present in the tea due to the fact that many other compounds either did not absorb within the pot or were lost during burial. (Reprinted from Eleanora A. Reber and Matthew T. Kerr, "The Persistence of Caffeine in Experimentally Produced Black Drink Residues." *Journal of Archaeological Science* 39[7]:2312–2319, 2015, with permission from Elsevier)

In sum, we determined that caffeine does last in archaeological residues and can be detected and analyzed using an untargeted analysis as long as chloroform/methanol extraction is used. An analyst can detect caffeine in pottery residues even if the analyst and archaeologist are not specifically looking for it. If possible, once caffeine is detected serendipitously, the analyst should adopt a targeted approach to the residues. This would involve a second, hot water extraction of the samples to maximize detection and sensitivity of all methylxanthines, including theobromine, and ideally analysis on an HPLC/MS.

NICOTINE, SCOPOLOMINE, AND OTHER POTENTIAL ALKALOID BIOMARKERS

Because alkaloid detection generally works better using targeted analyses, analysis of alkaloid biomarkers in archaeological residues is relatively rare. This is probably because determining which alkaloids to target in which types of vessels is not obvious.

Aside from methylxanthines, the most commonly targeted alkaloid biomarker is nicotine (Carmody et al. 2018; Echeverría et al. 2014; Rafferty 2002, 2016; Tushingham et al. 2018, and others). Nicotine is an obvious alkaloid to target for several reasons. Although nicotine is found in some plants that are not in *Nicotiana* (US Department of Agriculture, Agricultural Research Service 1992–2016), it is usually a reliable biomarker for various types of tobacco. Although nicotine has also been detected in ceramic vessels (Zagorevski and Loughmiller-Newman 2012), tobacco is often smoked, and so pipes can be selected for targeted sampling of nicotine by-products. Nicotine and its by-products can be extracted using chloroform and methanol, so that untargeted and targeted extraction can be combined. It can then be detected relatively easily using HPLC/MS or GC/MS, often using SIM mode. Further, when sampling pipes, there is often a visible residue on the interior of the bowl or pipestem, and so nondestructive sampling may be possible. Sean Rafferty pioneered the original technique for the targeted analysis of nicotine in 2002 (Rafferty 2002), but it has since been adopted by many different analysts and archaeologists.

Few other alkaloids provide such obvious cues for targeted analysis as nicotine, and none has been sampled as widely as caffeine, methylxanthines, and nicotine. In a few cases, however, alkaloids have been successfully targeted and detected. As briefly discussed in the note above, Adam King and colleagues targeted atropine and scopolomine, in a series of vessels, both ceramic and shell, from west Mexico, Mississippian, and Caddoan contexts curated at the Gilcrease Museum in Oklahoma. These compounds are the active constituents in the hallucinogenic plant *Datura stramonium*. They found atropine in 13 pottery vessels and 14 shell vessels (King et al. 2018). Several of the west Mexican samples bore decoration that appeared similar to the appearance of datura seed pods (King et al. 2018). Interestingly, the Mississippian and Caddoan vessels did not show motifs obviously based on datura. Instead, the motifs were previously identified as potentially symbolizing the Black Drink. In this case, burr sampling was used to minimize destruction to the rare complete vessels in the project (King et al. 2018).

Targeted analysis of atropine and scopolamine is best carried out using HPLC because of the high boiling point of both compounds, at 429°C for atropine and 460°C for scopolamine. Although derivatization permits the analysis of these compounds with GC/MS (Namera et al. 2002), targeted HPLC/MS is more sensitive.

Capsaicin, the biomarker for chili peppers, was targeted in pottery from Preclassic contexts in southern Mexico by Terry Powis and others (2013). In this case, the ceramics were burr sampled and originally targeted for caffeine and methylxanthines from cacao production, which were not found. As chili pepper is often used for flavoring chocolate beverages in Mexico, the analysts also targeted capsaicin, eventually discovering the compound in five samples and methylxanthines in none (Powis et al. 2013).

Many alkaloids must be detected using a targeted approach or are best approached in this way. This, of course, will mean that the analyst needs to know what they are looking for in order to tailor a targeted approach to the solubility and boiling point of the compound of interest. For this reason, relatively few alkaloids have yet been investigated using residue analysis, and only methylxanthines and nicotine are regularly targeted at the present time.

A Note on Residue Analysis on Nonceramic Artifacts

Generally, residue analysis can really only be performed on ceramics. This is because, as mentioned, the firing of ceramic items removes the natural lipids from the clay, so that only lipids produced and deposited after the production of the vessel can be absorbed. Almost every other type of artifact, from stone vessels to shell cups, contains lipids from before the production of the artifact (Buonasera 2007; Reber and el-Behaedi 2016). Distinguishing the prearchaeological lipids from those deposited by people is extremely difficult and is rarely possible. Biomarkers, and especially alkaloid biomarkers, are one of the few cases where nonceramic containers or artifacts can be sampled with some confidence. Assuming that there is no natural way for the biomarker of interest to absorb in or deposit on the nonceramic artifact, it therefore follows that sampling is possible. If the biomarker is detected on or in the nonceramic vessel, it must then derive from archaeological use. For this reason, stone pipes can be sampled for nicotine (Tushingham et al. 2018), and shell cups can be sampled for methylxanthines and atropine (King et al. 2018). Modern contamination, of course, can take place with nonceramic artifacts as well as with ceramics.

ALCOHOL AND OTHER NONALKALOID PSYCHOACTIVES

Although alcoholic beverages do not usually contain alkaloids (unless some are deliberately added), they are psychoactives that are important in archaeological interpretation of pottery use. Pure alcohol is impossible to detect in residues, as ethanol is a small, volatile compound that easily reacts with other compounds and evaporates out of residues. Detection of alcoholic beverages therefore depends on the detection of biomarkers for the substances fermented, or of by-products of the fermentation process.

Wine is the alcoholic beverage most commonly detected in residues. Similarly to the detection of fish, discussed in chapter 7, there is no specific biomarker for the presence of wine. Instead, a series of indicators are most effectively used along with the presence of supporting ethnobotanical and archaeological analysis. Tartaric acid is one of the most commonly used biomarkers for wine (Guasch-Jané et al. 2006; McGovern et al. 1996, 2017, and others), but it is also found in a variety of other fruit, is prone to washing out of residues in nonarid regions, and may suffer from contamination during deposition (Drieu et al. 2020; Roffet-Salque et al. 2017). It is a useful indicator for wine in residues, especially from arid regions with no nongrape fruits in the area containing large amounts of tartaric acid, and especially with archaeological evidence for grape processing (Drieu et al. 2020). Syringic acid has been used to indicate red wine in residues (Barnard et al. 2011), but the extraction methodology must be carefully calibrated to ensure that the syringic acid is deriving from pigments in grapes rather than from other foods that contain the compound (Drieu et al. 2020). There are a variety of short compounds often used as indicators for the presence of fermentation, including malic acid, succinic acid, citric acid, and lactic acid (Blanco-Zubiaguirre et al. 2019; McGovern et al. 2017, and others). These compounds, however, occur in a range of fruits, fungi, and soils and can easily wash out of residues. They are most effective when used to support other indicators for wine production (Drieu et al. 2020). Wine detection, in short, is more difficult than it originally seems and is highly dependent on careful contamination controls and the depositional environment at the site in question. The more indicators for wine that are present in a given residue, the more convincing the interpretation.

Calcium oxalate has been proposed as a biomarker for brewing beer (Michel et al. 1993; McGovern et al. 1999). This water-insoluble compound is produced as part of the brewing process and has occasionally been detected as visible, yellowish encrustations on pottery (Michel et al. 1993; McGovern et al. 1999) or in visible residues (Wang et al. 2016). This compound is usually used in concert with starch grain, phytolith, and archaeological analysis to detect beer production. As calcium oxalate occurs in a range of plants, it cannot act as a biomarker without support from other lines of evidence. Recent publications have suggested a range of compounds as possible markers for fermentation of grains in absorbed residues (Perruchini et al. 2018). These are intriguing, but further investigation is probably needed to determine whether these compounds could derive from breakdown of a non-fermentation-based archaeological residue. At the present time, identification of beer in residues is still under development.

Perhaps the most useful takeaway from the analysis of alcoholic beverages in residues is that detection of alkaloid biomarkers is easier and more definitive than that of alcohol and fermented beverages. Alkaloids preserve better in pottery, are usually less prone to issues of contamination, and are more distinctive to their sources. A discussion of hopanoid triterpenoids as biomarkers for pulque production in Mexico and cannabinols as biomarkers for marijuana use is included in chapter 5.

CONTAMINATION ISSUES WITH ALKALOID BIOMARKERS

Although most alkaloids do not have a strong potential for contamination in residues, contamination can still sometimes occur. In general, alkaloid contamination is a problem in the following circumstances:

(1) The alkaloid is in common enough modern use that it might somehow touch an archaeological pottery sample.

(2) The alkaloid either aerosolizes easily without chemical transformation or is commonly carried on people's fingers.

(3) The alkaloid easily absorbs within a pottery surface once transported there.

Clearly, therefore, alkaloids such as caffeine and nicotine are the contaminants to worry about the most. Unless there is record of people regularly smoking psychoactive drugs either on-site during excavations or in a museum area, rarer psychoactives are not usually a concern for contamination.

Caffeine in particular can be a dangerous and persistent contaminant, however. Caffeine-containing beverages are commonly drunk during excavation and laboratory processing and are often actually made in museum artifact storage areas. Caffeine itself has a high boiling point, but easily aerosolizes and is transported on dust. Once transported, it can be deposited on artifacts in open storage. This was demonstrated by Dorothy Washburn and colleagues, when they sampled a variety of swabs from open museum storage (Washburn et al. 2014). The study did determine a floor concentration at which the authors thought that caffeine in a residue probably derived from archaeological use. The most important result of the study, however, was an appreciation of the sheer amount of caffeine that had built up on open museum storage shelves over the years. This conclusion has been supported by other research suggesting that, in some cases, contamination levels may overwhelm methylxanthines from archaeological uses (King et al. 2017). As a result, both archaeologists and analysts became much more cautious about caffeine contamination during artifact storage and processing.

For this reason, it's important to carefully consider the storage history of any item analyzed for caffeine (and possibly nicotine) presence. Items stored for long terms in open-shelved storage are at real risk of contamination, and shelves surrounding the artifact should be swabbed for testing at the same time that the artifact is chosen. Swabbing methodology should also be discussed ahead of time with the residue analyst, and clean swabbing supplies used. Any items regularly stored in the open, especially around the museum, might be at risk of caffeine contamination themselves. In addition, items commonly used by curators to swab surfaces may not be clean in residue terms. That is, they might themselves contain lipids that would complicate the detection of caffeine in the residue lab. The residue analyst should be able to produce a good supply of solvent-washed swabbing items and pure solvents for use in contamination sampling.

Artifacts stored in closed contexts (boxes, bags, and such) will probably have escaped this source of contamination and are the optimal type of artifact to sample for caffeine and nicotine. It is also important to consider laboratory contamination when performing this type of analysis.

Case Study: Caffeine and Contamination of a Shell Cup from Florida

In 2016, Nancy White, of the University of South Florida, submitted a large *Busycon sinestra* (lightning whelk) shell cup for targeted caffeine analysis. The analysis was unconventional, as it was not pottery, but there is evidence that compounds can be preserved on vacuoles in the surface of shell cups (King et al. 2018; Reber and el-Behaedi 2016). Caffeine is not known to naturally occur in any shellfish, including *Busycon sinestra*, and so natural shellfish lipids would not be a problem in the study. We were strictly looking for caffeine. In addition, the cup had been stored in a cardboard box since its original excavation, intact, with the interior soil present. In terms of its curation and storage, the artifact was perfect. Then-undergraduate Didi el-Behaedi and I took on the project.

After combing through the literature, we contacted Ralph Mead in the Department of Chemistry and Biochemistry at UNCW about using his HPLC/MS. He generously agreed, and we sampled the shell nondestructively, using three 5-minute hot-water rinses, shown in Figure 4.10, that were combined and freeze-dried, as described in Washburn et al. (2014). The fluff was reconstituted and analyzed on the HPLC/MS along with a series of external standards (to quantify the amounts of methylxanthines) and solvent blanks. We were ecstatic when the shell cup sample contained caffeine! Our disappointment was extreme when the solvent blank also contained caffeine. Somehow, caffeine was contaminating our laboratory procedure, despite all our precautions and the fact that caffeine-containing beverages are forbidden in the Residue Lab. After careful reconstruction and self-examination, we determined that the likeliest source of the contamination was either from the production of external standards in the Residue Lab or in the freeze-drier.

When external standards are prepared, small amounts of known caffeine and other methylxanthines are dissolved within the solvent of interest—in this case, water. It is possible that a small flake of caffeine somehow aerosolized during the measuring process and contaminated work surfaces, even though we cleaned them carefully. It is also possible that freeze-drying the

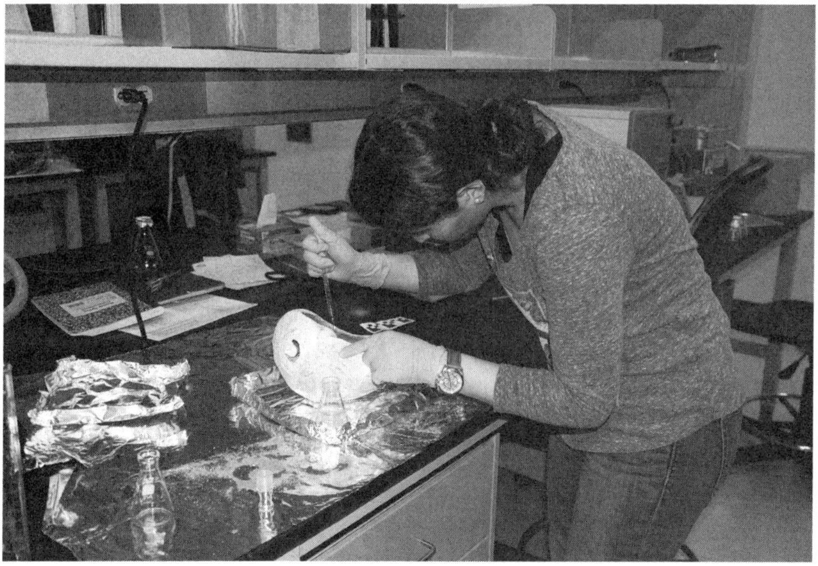

Figure 4.10. Didi el-Behaedi conducting hot water extraction on the *Busycon* shell cup submitted to the lab for caffeine detection by Nancy White. The surface is lined with aluminum foil to minimize surface contamination from the counter to the cup. (Courtesy Didi el-Behaedi and Nancy White)

archaeological samples and blanks at the same time in the freeze-dryer led to some sort of cross-contamination from the archaeological samples to the blank, although this seems unlikely. Whatever the mechanism of contamination, we could not guarantee that the caffeine found in the shell cup derived from archaeological use, as we knew there was a contamination source in the lab. The cup could not be resampled, as the hot water extraction would have likely removed all the easily available methylxanthines. The results of the shell cup analysis were therefore never published.

As an archaeologist the key takeaway from this case study is to remember the sheer contamination potential of caffeine in laboratories as well as in artifact storage facilities (King et al. 2017; Washburn et al. 2014). When interested in caffeine and methylxanthine analysis, be certain that your analyst has adequate controls in place to control for potential contamination. After this incident, our lab now makes the external standards after all archaeological samples have been extracted, sealed, and stored in a separate room, and we freeze-dry the archaeological samples and blanks in different containers. The external standards are made in a different part of the lab than the extraction area, and this area is scrubbed down three times with hot water and soap after the standards are made. We scrub all surfaces down with hot water and soap in between all steps of the caffeine extraction projects. And, of course, caffeinated beverages are still never permitted in the lab.

A Note on Caffeine in the Field

Given the contamination potential of caffeine, it is reasonable for an excavator to wonder whether caffeinated beverages should be permitted in the field if they know that some of the sherds under excavation will later be analyzed for caffeine or methylxanthine content. In the interests of crew morale, I would argue yes, but with caveats. As long as the caffeinated beverages are kept out of the excavation areas, and the site is in open air, methylxanthines from beverages are unlikely to aerosolize and recrystallize into the units. Instead, any aerosolized methylxanthines are likely to blow off-site, or be so diluted by the fresh air that their contamination potential is greatly diminished. In addition, caffeinated beverages tend to aerosolize more when kept heated (as in a coffee pot or percolator) or in open cups. So beverages in a covered travel mug should be fairly safe.

If, however, a cup of caffeinated beverage is brought into a unit during excavation, the calculation changes. If any of it spills, then methylxanthine analysis is unreliable for that unit. The situation is similar for field laboratories. The excavation or field processing crew might want to take a closer look at some particularly interesting artifacts—if they are holding caffeinated beverages and potentially spilling them, contamination is a real danger. A coffee pot in a laboratory area, especially if artifacts are regularly left on trays for any length of time, is a potential source of contamination. It should be avoided. If you know that future caffeine screening is important to your research design, it will be useful to make sure your excavation and lab crew know to keep caffeinated beverages away from artifacts of interest, and to keep the coffee maker in a different room from artifact storage.

ALKALOIDS AND PSYCHOACTIVES: AN EXCITING FIELD OF STUDY

Every alkaloid found in a residue represents either a category biomarker or a resource-specific biomarker. It also almost certainly holds valuable information about human use and control of the physiochemical effect caused by the particular compound. Often, these physiochemical effects have a cultural meaning, as in the use of caffeine-containing Black Drink as a purifying agent in many southeastern Indian cultures (Hudson 2004). Alkaloids therefore represent unusually large and specific amounts of information when identified in a residue. They tell us in part the contents of a vessel or pipe, the physical effects at least part of the contents produced, and possibly ritual or culturally meaningful behaviors linked to the item containing the residue. Because of the great variety of chemical forms of alkaloids, however, they often need to undergo some sort of targeted analysis, sometimes including specialized extraction techniques. The archaeologist is going to be the best source of information on what alkaloids might be used in specific types of pottery at an archaeological site; the residue analyst will not usually have this information. So, when planning research designs for a site, keep in mind the possibility of alkaloids in some pottery types, including pipes and more conventional vessels.

This chapter has focused on a few, well-known, and commonly analyzed alkaloids and psychoactives. Given the wide range and physicochemical activity of this compound class, however, this type of analysis will continue to develop in the future. Further analysis of a wide range of targeted alkaloids and psychoactives should be possible, limited only by the communication between archaeologist and residue analyst.

5

PLANT RESINS

Alkaloid biomarkers, as described in the previous chapter, are seldom found in untargeted residue analysis. Some types of nonalkaloid biomarkers, however, are persistent in residues, found in untargeted analyses, and permit interesting interpretations. These biomarkers can be found serendipitously and without additional targeted extraction and analysis methodologies. Aside from sterols, the most ubiquitous of these are terpenoid biomarkers for tree and plant resins. Although most of these compounds are not psychoactive, they do sometimes have medicinal and ritual use in the form of incense, wound care, and teas. As such, the terpenoids are an unusually informative and helpful group of compounds found in residues.

Some important plant resins are composed of nonterpenoid biomarkers, most notably the phenolic resins found in a variety of sources, including the sweetgum tree (*Liquidambar styraciflua*) and cannabis (*Cannabis* spp.). Such phenolic compounds are also stable biomarkers for plant resins that can be extracted and identified in untargeted analysis. Like terpenoid resins, phenolic resins were used in a variety of important ways archaeologically that make their identification in a residue informative on several different levels. When a plant resin, such as that from *Cannabis* spp., has psychoactive properties, interpretation of its psychoactive use is similar to those of alkaloid psychoactives, discussed in the previous chapter. The extraction methodology, however, may be untargeted and the detection process is identical to that of nonpsychoactive resins.

This chapter explores the different types of plant resins that can be detected in residues, including diterpenoids, triterpenoids, and phenolic compounds. All are persistent in residues, and some of these compounds are resource-specific biomarkers or category biomarkers. In most cases, however, these compounds are found in a wide range of plants and require cooperation between the archaeologist, paleoethnobotanist, and residue analyst to obtain the most specific interpretation of the source of the resin.

A Note on Distinguishing Resins

When fresh, pure resins may be molded into shapes, used as glue, and burned as incense. An important point to keep in mind is that although diterpenoid, triterpenoid, and phenolic resins look different to a residue analyst, their physical properties are very similar. All look similar, feel similar, and smell aromatic and are often produced by plants and help heal wounds or prevent insect infestations. As a result, all have potential medicinal benefits. To the ancient people, therefore, these types of resins may have been used almost interchangeably. For example, Mesoamerican copal appears to refer to any resin that may be used as an incense or adhesive, including diterpenoid conifer resins and triterpenoid resins from genus Bursera (Stacey et al. 2006).

Occasionally, resins can be produced by insects, most famously lac, a resinous secretion produced by a variety of insects in Asia (Langenheim 2003; Stacey et al. 1998). Resin-producing insects are also present in the Great Basin desert, and lac resins can be found in western North America (Stacey et al. 1998). In general, lac resins tend to be a mixture of wax and terpenoid components apparently derived from the diet of the resin-producing insects (Stacey et al. 1998). As a result, insect-based lac resins may potentially be confused with plant-based terpenoid resins in a residue, although the presence of unique types of alkanols typical of insects should help distinguish the two resin sources (Stacey et al. 1998).

BACKGROUND ON TERPENOIDS AND TERPENOID RESINS

Chapter 2 provided some background information on terpenoids. More specifically, terpenes are a variable class of compounds made up of a single building block—the isoprene (or terpene) group. This is simply four carbons with a single branch and two double bonds, as shown in Figure 5.1. These compounds can snap together, like Tinkertoys, into a wide number of chemical possibilities including complex ring shapes. Compounds made of isoprene groups with only carbon and hydrogen are terpenes; those with additional components, such as oxygen, are terpenoids. The lighter of these compounds (monoterpenoids and sesquiterpenoids) are often important in the taste and smell of herbs, fruits, and aromatic plants, but they are too volatile to survive in archaeological residues. Because of the number of potential combinations that can be constructed from the isoprene group, terpenoids

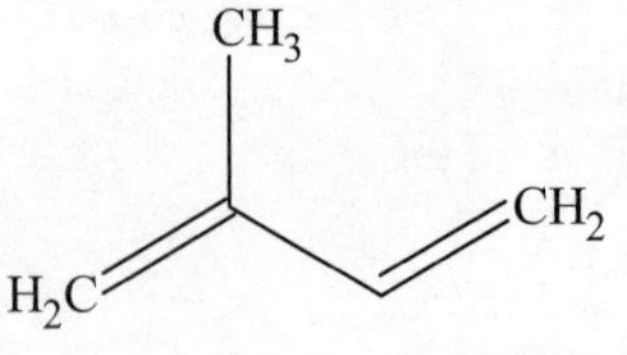

Figure 5.1. Structure of a single isoprene group. (Eleanora A. Reber)

include a large number of compounds, including all sterols. In a way, then, sterols are terpenoid biomarkers that can be found in both plants and animals, but because they are a specialized class of compound, they tend to be conceptualized as separate from the broader universe of terpenoids.

The smallest terpene compounds that are likely to survive in residues are diterpenes, which are composed of four isoprene groups. All terpenes are described using the number of rings present in the structure. For example, the isopimaric acid and dehydroabietic acid shown in Figure 5.2 are tricyclic diterpenoids (made up of four isoprene groups, containing three rings and oxygen), whereas the communic acid in that figure is a bicyclic diterpenoid. Diterpenoids are distinctive compounds with a variety of different skeletal structures, all of which are present only in plants, and some of which are present only in conifers.

Triterpenoids are made up of six isoprene groups. Pentacyclic triterpenoids, like the α-amyrin in Figure 5.2, are present only in plants, and some of these can serve as biomarkers for either a class of plant, such as a nonconifer tree, or even a specific type of tree. Betulin, as mentioned, is a specific pentacyclic triterpenoid found primarily in birch trees.

In general, the more rings present and the more conjugated double bonds present within the rings, the more stable a compound is in an archaeological residue. It should be clear from Figure 5.2, then, that cyclic diterpenoids and triterpenoids are very stable compounds. They tend not to easily hydrolyze or oxidize, except into even more stable terpenoids. They are not usually water soluble, even in hot water, but when they do become part of a residue, they are remarkably stable. This may occur in several ways. First, by deliberate application of resin or resinous plant material to the pot surface. Second, by cooking or heating terpenoid-containing resources, which melts the terpenoids and deposits them on the interior surface of a vessel. Third, by soot deposition during pottery firing or cooking.

Because of their stability and nonsolubility in water, terpenoids last for a very long time in a residue when deposited. Further, their unique properties, including waterproofing, taste, smell, and potential medicinal uses, mean that they are deposited in residues surprisingly often, either deliberately or as a side effect of pottery firing or processing. The problem then becomes

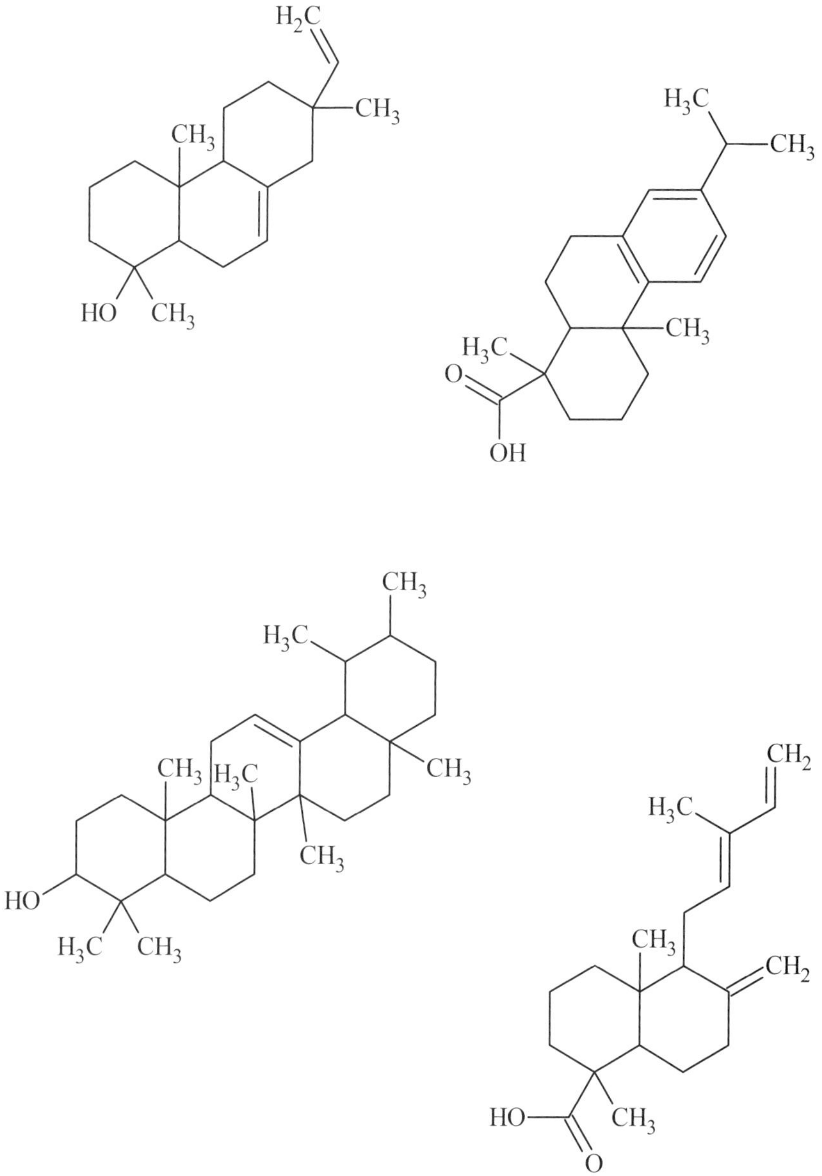

Figure 5.2. *Top left*, isopimaric acid, a pimarane tricyclic diterpenoid found in conifer resins and often used in residue interpretation. *Top right*, dehydroabietic acid, an abietiane tricyclic diterpenoid, commonly produced by oxidation of abietane diterpenoids. *Bottom left*, α-amyrin, a pentacyclic triterpenoid. *Bottom right*, communic acid, a labdane bicyclic diterpenoid. (Eleanora A. Reber)

interpreting why (and sometimes how) the terpenoids were deposited in the residue. This chapter illustrates some ways to address these puzzles through a variety of case studies involving terpenoid resins in residues.

Diterpenoid Biomarkers

Diterpenoids are a common and helpful category biomarker. They are easily detected in untargeted analysis, and two of the major types of diterpenoids are found primarily in conifer resins. These are pimarane and abietane diterpenoids (see Figure 5.2), terms that refer to the overall structure of the compounds. Both are tricyclic, and members of each type can convert into other compounds of the same type in the presence of heat, oxygen, hydrolysis, and such (Mills and White 1977). Pimarane diterpenoids cannot convert into abietanes, however, or vice versa.

Labdane diterpenoids, such as communic acid (see Figure 5.2), are bicylic and are found in a wide range of plant resins in both conifers and nonconifer resins. Communic acid is an important component in resins from the conifer family Cupressaceae (including cedars and junipers) as well as in many nonconifer resins, such as those from Leguminosae. This family includes members of genus *Hymenea*, well known for producing diterpenoid resins including Brazilian copal (Langenheim 2003; Mills and White 1977). Resins from Leguminosae look similar to those from many conifers and are also composed largely of diterpenoids. Chemically, however, the resins may be easily distinguished, because there are no abietane or pimarane diterpenoids in the resins from Leguminosae, which are composed of a range of other diterpenoids and lighter compounds, depending on species (Langenheim 2003; Mills and White 1977). Labdane diterpenoids may also co-occur with triterpenoids in nonconifer resins (Langenheim 2003; Li et al. 2014; Mills and White 1977)

Kaurane diterpenoids are tetracyclic (four rings) and are also found in both conifer and nonconifer resins, although they are most abundant in nonconifer resins (García et al. 2007; Langenheim 2003).

Case Study: Pot Sealing and Diterpenoids

The second residue project I worked on at UNC Wilmington was a series of residues, both absorbed and visible, from sites in New York State. John Hart, of the New York State Museum, is interested in a number of issues regarding changing diet and pottery in New York State and the northeastern United States in general. Among these issues are the adoption of maize and the potential adoption of pottery in the region to help process the abundant nut resources in the area. Over the course of several years, we looked at 17 visible residues from the area and 12 absorbed residues, with eight sherds sampled for both absorbed and visible residues (Hart et al. 2008; Reber and Hart 2008a, b).

Somewhat to my surprise, we found diterpenoid biomarkers for conifer resins in 10 of the 12 absorbed residues and 11 of the 17 visible ones. Three sherds contained diterpenoids in both the absorbed and the visible residues, with two sherds containing diterpenoids in the absorbed residues but not the visible ones, and one sample contained diterpenoids in the visible residue but not the absorbed (Reber and Hart 2008a). These compounds surprised us, largely because at the time we were looking for evidence of broader dietary change. Conifer resins hadn't really registered as a major resource in the area, and yet small amounts of oxidized diterpenoids appeared in 18 of 25 samples using two different methodologies.

The most common diterpenoid identified was the very stable compound dehydroabietic acid, which is the oxidative by-product of four commonly found abietane diterpenoids (Mills and White 1977). However, if a residue includes abietane diterpenoids, it can probably be safely identified as containing some sort of conifer resin (Mills and White 1977). Although pine resins (*Pinaceae*) are among the types of conifer that have the most abietane diterpenoids, as well as the similar pimarane diterpenoids, other conifers also have these compounds, including the native tamarack (*Larix larcena*) and firs (*Abies balsamea*) (Mills and White 1977). Given the number of native conifers in New York, the diterpenoids in the residues could have derived from several species of pines and spruces, tamarack, and/or fir. Given the degree of oxidation of the resins, it could have been several of them. Figure 5.3 shows an example of the presence of dehydroabietic acid and the small quantity present in many of these residues.

Naturally, with any group of residue studies it's dangerous to assume that all the diterpenoid resins came from the same source. The study included a wide range of sherds from several different sites and time periods in the region, and different conifer resins could have been used at different times and different places. In fact, one of the residues, from the Vinette site, contained a sequence of diterpenoids that looked completely unlike all the other compounds in the study. It had been used to process a different resin from the others. Determining the origin of the residues was also complicated by the fact that most of the diterpenoids were present in small amounts in an oxidized form.

The small amount and highly oxidized condition of the diterpenoids did supply a potential clue to the way in which the resin was used in the pottery. The pottery didn't seem to have been used to process conifer resin or turpentine—such abundant resin would have provided more diterpenoids and almost certainly a wider range of compounds. Further, when pine resins are

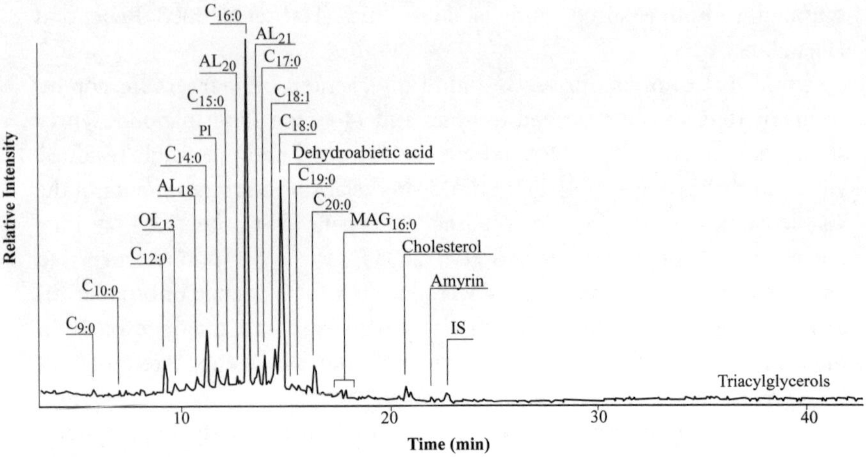

Figure 5.3. TLE of an absorbed residue extracted from Kipp Island 41119-5 using chloroform/methanol extraction. Note the small amount of dehydroabietic acid found in the residue as well as a variety of compounds from resources other than conifer resins, including the triterpenoid amyrin. This compound indicates resinous compounds from nonconifer plants. (Eleanora A. Reber and John P. Hart, "Pine Resins and Pottery Sealing: Analysis of Absorbed and Visible Pottery Residues from Central New York State." *Archaeometry* 50[6]:999–1017, 2008. John Wiley and Sons, CC-BY)

heated for a length of time (as one usually does to process pine pitch), a further oxidative step takes place, producing the process biomarker retene. No retene was found in any of the residues. In addition, all of the residues included resin as one of several different components, including meat-based compounds and probable plant foods (based on cholesterol and alkanols). The residues, in general, looked similar to standard cooking residues, except with the addition of a small amount of oxidized conifer resin, as shown in Figure 5.3. This residue contained components deriving from meat (the cholesterol) and nonconifer plant resins, as indicated by amyrin. This is a pentacyclic triterpenoid not found in conifers but present in a range of nonconiferous plants (US Department of Agriculture, Agricultural Research Service 1992–2016).

The ethnographic literature provided some possible uses for coniferous resources. The resin, needles, or bark could have been used as a medicine, the needles could have been made into a tea, pots could have been waterproofed or sealed with it, and it could have been used as a flavoring. As mentioned, processing pure pine pitch or sap seemed unlikely. This left the other possible uses. Some form of medicine was possible—conifer sap has antimicrobial properties (Sipponen 2013), and indigenous Americans used pine sap as a sort of bandage over wounds (Moerman 1998). They also made a tea known to cure scurvy from the needles of a conifer, known to the Iroquois as Annedda, whose botanical identity is presently unclear. *Abies balsamea* (balsam fir), *Thuja occidentalis* (arborvitae), *Pinus strobus* (white pine), or any of several other varieties of conifers native to New York and Canada contain sufficient vitamin C and potentially helpful amino acids to relieve scurvy (Durzan 2009). In addition, a variety of medicinal uses are recorded for decoctions of conifer trees for many different conditions, including colds, rheumatism, and boils (Moerman 1998). On consideration, however, it seemed unlikely that so many of the pots in the study would have been used for processing small amounts of tea or decoctions in vessels that were normally used for food. In addition, the diterpenoids present in the residues were oxidized and minor components in the residue, suggesting that pure resin or large amounts of tea were not processed in the vessels.

That left flavoring and pot sealing as the most likely hypotheses to explain the widespread presence of small amounts of conifer resin in residues from New York State. The resin is edible in small amounts and imparts a piney flavor that would have added interest to many dishes. Spruce beer was a popular soft drink for some years, and recipes for pine needle salad dressing and baked goods made from pine bark used as a flour substitute are available on the internet (Adamant 2018, and others).

The most likely option, after much consideration, seemed to be pot sealing. Porosity is a common problem for potters using unglazed pots—liquid contents can easily seep through the pottery, causing real complications during the cooking process. One solution for this problem lies in waterproofing or sealing the vessels, at least partially. This sealing keeps liquid pot contents from leaking through the ceramic, at least for long enough to allow the pot to reach temperatures that allow boiling or simmering. Many different substances have been used around the world to seal a pot, and in fact, almost any nontoxic substance will work as long as it somehow blocks the pores of a pot. Sealant may be applied to either the interior or the exterior of the pottery vessels. Ethnographically, the Pamunkey of Virginia sealed their pots by smearing the interior with bear grease (Fewkes 1944). The Kalinga, in the Philippines, are known to have applied pine resin to their unglazed cooking vessels and water jars (Longacre 1981; Longacre and Taylor 2015). A wide variety of substances, from milk to resins to pig blood, has been applied to either the interior or exterior of pots as sealants—and in some cases, both (Rice 2015:177–178). Given the abundance of conifers throughout New York, and the usefulness of pot sealing, it seemed to be a realistic possibility that conifer resin was used in this way in the area.

In the absence of definitive evidence, we were forced to rely on logic to provide some probable markers of pot sealing in residues. If pottery sealing is in common use, the sealant should be present in most pots from a given culture, but only in relatively small amounts. If the sealant provides a biomarker, the biomarkers should be fairly heavily oxidized, as the sealant would be added to the pot soon after production. It would then constantly undergo oxidation and hydrolysis throughout the rest of the pot's use-lifetime. The residues we investigated from New York State did contain these markers, and we found it most likely that the resins in the New York State residues derived from pot sealing. This was somewhat supported by the fact that one of the residues from New York contained biomarkers for bitumen, which can be used as a pot sealant but is not commonly used as a flavoring agent (Reber and Hart 2008a).

Clearly, there are many uses for conifer resins. That said, given the ubiquitous presence of conifer resin in the samples, the presence of residues from other foods, and the highly oxidized condition of the resins, sealing seemed to be the most likely explanation for the diterpenoids in New York pottery.

To arrive at these conclusions, we used a mix of techniques and pieces of evidence—the pottery residue analysis, obviously, but also the paleobotany of New York State, ethnography from a variety of sources, and general archaeological knowledge about the people who made the pots we were sampling. This mixture of archaeometric and archaeological problem solving is typical of many pottery residue problems.

Pottery Firing and Pottery Sealing

Later research has suggested that it's actually possible for diterpenoid resins to enter pottery residues during the firing process. In an experimental study carried out by Matthew T. Kerr (of the University of Tennessee Knoxville), Helen Whelton, Richard P. Evershed (both of the University of Bristol), and me, we fired tiles from several different clays in a variety of conditions (Reber et al. 2019). The idea was to determine whether any lipids remained from the original soils in low-fired pots, which are the dominant type of pottery in the southeastern United States. For tiles fired in a furnace, no lipids were present. Tiles fired in an open pit with a mixture of oak and pine, as shown in Figure 5.4, did contain large amounts of abietane and diterpenoid biomarkers that were typical of pine resins, even with the exterior surface of the sherds removed during sampling. Clearly, firing the pots in the open pits allowed diterpenoids from the resin in the firewood to aerosolize in the soot, which was then deposited on the pottery. The diterpenoid compounds absorbed within the surface of the pot, producing a residue that appeared to mirror a deliberate sealing of the pot (Reber et al. 2019). In fact, it is possible

Figure 5.4. Experimental pottery firing of three different types of clay in an open pit using a mixture of oak and pine wood. The pottery was balanced on three equidistant fire bricks placed under the fire, with one of the bricks also holding the thermocouple of the pyrometer, as seen toward the front of the fire. (Eleanora A. Reber)

that people relying on pit firing would have deliberately used resinous woods during firing. The soot from the fire deposited on the vessels would have acted like a pot sealant, but without the additional work required by applying a sealant separately.

Distinguishing between resin sealing produced through soot during firing and resin sealing produced by deliberate application of resins is quite difficult using residue analysis alone. The only possible way to do this would be to look for biomarkers of soot within the residue. The most well-known of these biomarkers are the highly ring-based unsaturated compounds known as polyaromatic hydrocarbons (PAH). If PAHs are found in a residue, as well as diterpenoid biomarkers, then the presence of soot in the residue would be established. It would then be possible to argue that the diterpenoid biomarkers also originated from the soot, but it couldn't be directly determined. Further, PAHs seem not to absorb within pottery as easily as diterpenoids. In the experimental study, we did not identify large amounts of PAHs in the residues, even though we knew the pots were sooted, and we all noted the strong smoky smell of the pottery during production and sampling.

Case Study: Same Compound, Different Tree?

As so many conifer resins have overlapping diterpenoids, and many of these diterpenoids oxidize into a relatively small group of stable compounds, it is possible, and even likely, for resins from different conifer species to look similar in a residue. For conifer resins, or for other resinous compounds that occur in many different plants, communication between the analyst, the archaeologist, and an ethnobotanist will be crucial in producing the most accurate interpretation.

For example, in two separate projects in the midwestern United States, I have identified conifer resins that look chemically similar to the New York State residues discussed. One, analyzing 25 absorbed residues from Hopewellian sites in Kansas, found oxidized abietane diterpenoids in the form of dehydroabietic acid or 7-oxo-dehydroabietic acid in 17 of the samples. In the other, analysis of 31 samples from the George Reeves site (11S650) identified 16 residues containing a variety of abietane and pimarane diterpenoids.

Figure 5.5 shows a close-up of one of the residues extracted from a sherd from the George Reeves site—a limestone-tempered, cordmarked sherd. It shows abundant residue as well as a small amount of dehydroabietic acid. It therefore looks superficially similar to Figure 5.3, which also showed dehydroabietic acid and a range of other resources that did not derive from resin.

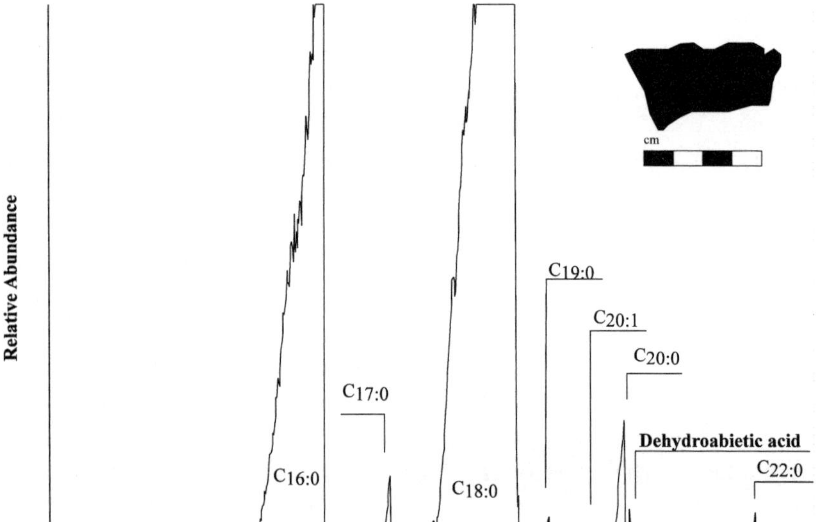

Figure 5.5. Close-up of portion of the total acidic hydrolysis fraction extracted from a limestone-tempered cordmarked sherd from the George Reeves site, shown in silhouette. The residue contains dehydroabietic acid, indicating an oxidized conifer resin, along with abundant fatty acids from the sherd. (Eleanora A. Reber)

The interpretation of this residue, however, is different from the Kipp Island residue shown in Figure 5.3 because the ethnobotany of the Mississippi valley is different from that of New York State.

There are many possible conifers in New York State, any one of which could have supplied the dehydroabietic acid in the residues. In the Mississippi Valley near George Reeves, however, conifers are quite rare. Indeed, the only native conifer in the site region now is the eastern redcedar (*Juniperus virginiana*), shown in Figure 5.6. Although a conifer, redcedar is not a pine but is from a genus generally believed to contain primarily pimarane and labdane diterpenoids (Mills and White 1977). Further investigation of resins from the eastern redcedar, however, determined that both pimarane and abietane diterpenoids are present in this particular member of *Juniperus*. The dehydroabietic acid in the residues could therefore have derived from redcedar.

Further investigation of the ethnobotany of the area, after consultation with the generous and helpful ethnobotanist Neal Lopinot, suggested that the most likely source of the resin found in both the George Reeves sherds

Figure 5.6. Eastern redcedar tree *Juniperus virginiana*, showing the typically dramatic contrast in wood color shown where the limb was removed. (Eleanora A. Reber)

and the Kansas Hopewellian sherds was eastern redcedar. There were several other conifers that once existed in the general region of both groups of sites, but in both cases, eastern redcedar was the only native conifer present in the site area and had the resin that most closely matched the (admittedly highly oxidized) resins present in the residues.

This allows a more precise identification of the resources processed in vessels at these midwestern sites than was possible in the New York State sites. It also raises some interesting possibilities in terms of human behavior. Eastern redcedar resin has all the standard medicinal and adhesive uses of a resin described and additionally is known to have been of considerable ritual importance to the native people of the Midwest, especially in the Mississippian and possibly Emergent Mississippian periods (Simon 2002). The color of the wood is a contrasting red and white, shown in Figure 5.6, both ritually important colors in many indigenous American cultures. The wood repels insects (due to the terpenoids in the tree resin), making it a metaphor for protection and preservation. The ritual importance of this tree has been discussed at greater length in other publications (Mohlenbrock 1986; Simon 2002). The presence of probable redcedar resin, therefore, could potentially reflect a ritual use of redcedar resin, needles, or bark. Such a ritual use could be superimposed over a medicinal or practical use of the resin, as in vessel sealing, or be used by itself, as a purely ritual application of the tree. In any case, redcedar wood would probably not have been used unthinkingly to fire pottery; the wood is rare and symbolic enough in the area that its use would have been specifically chosen by the people who produced or used the pottery.

A Note on the Difference between a Pure Resin and Resinous Plants

A pure resin is the sticky sap that exudes from a tree or other plant and can be used as a glue, an incense, a chewing gum, or a coating. Resins can usually be produced by cutting the bark or stem of a resinous plant and waiting for the sap to drip out of the wound. In some cases, as in Figure 5.7, resins may exude from a healthy plant to protect it. Resins are usually made of terpenoids or phenolics, with a few exceptions.

Resinous plants are plants that contain terpenoid or phenolic compounds in their sap, but they do not all exude resins when wounded. For example, many plants with a heavy herbal scent and taste contain small amounts of terpenoids or phenolics. These plants do not exude resins when wounded, but when made into a tea or decoction, terpenoids or phenolics are extracted from the plant and could enter a pottery residue. A residue that included these compounds would be described as deriving from a resinous plant, even if a lump of pure resin was not cooked in the

Figure 5.7. Resin exuding from a bristlecone pine, *Pinus longaeva*, taken at the Patriarch Grove, Ancient Bristlecone Forest in California, 2012. (Courtesy James C. Reber and Eleanora A. Reber)

pot. A process such as making a pine needle tea, for example, might produce a residue that looked chemically similar to one produced by adding pure pine resin as a flavoring or sealant. So plants that exude resin are always resinous, but resinous plants do not always exude resin. This is why your analyst may describe a residue as containing components from resinous plants, rather than specifically mentioning resin, unless sampling actual pieces of resin. Alternatively, the analyst may refer to terpenoid or phenolic compounds as denoting plant resins, when they do not necessarily mean pure resin, but rather resins potentially extracted through plant processing.

Triterpenoids

Triterpenoids are also useful biomarkers and are equally stable and persistent in pottery residues as diterpenoids. Triterpenoids are present in resins from many nonconifer trees and in nonconiferous resinous plants. As mentioned, triterpenoids may co-occur with labdane or kaurane diterpenoids in nonconiferous plants (Langenheim 2003; Li et al. 2014; Mills and White 1977). Unlike diterpenoids, there is no one specific type or structure of triterpenoid that can be matched to a particular family or group of trees or other plants.

However, several individual triterpenoids may be used as species or process biomarkers. The best known of these is probably betulin, mentioned in chapter 2, which is commonly used as a biomarker for birch bark or birch tar (Charters et al. 1993b; Evershed 2008b; Urem-Kotsou et al. 2002, and others).

Triterpenoids are usually either tetracyclic or pentacyclic. They may be found in pure resins, perhaps most notably from the sap of the *Protium copal* (copal tree) and other resinous members of Burseraceae. Triterpenoids are also often found in resinous plants that do not exude resin, as described in the previous note.

Several triterpenoids are found in a range of nonconiferous plants. Among these are amyrin, shown in Figure 5.2, and lupeol, a triterpenoid alcohol. These are often found in residues, although not at as commonly as abietane and pimarane diterpenoids, at least in our laboratory. The compounds are distinctive but found in a wide range of resinous plants, from trees (including junipers) to other plants (including squash flowers, sunflowers, dandelions, and wild grapes, among many others). Ursolic acid was discussed in chapter 4 when attempting to distinguish caffeine deriving from cocoa from caffeine deriving from yaupon or dahoon holly. It is commonly found in plant leaves, including yaupon holly, dahoon holly, and grape. Similar to many other triterpenoid biomarkers, it therefore acts primarily as a category biomarker for plants.

One important triterpenoid biomarker is miliacin, a compound that is present in *Paniceum miliaceum* (broomcorn millet). Miliacin has been used as a biomarker for the presence of broomcorn millet in soils (Courel et al. 2017; Jacob et al. 2005). After experimental studies established that the compound was absorbed within pottery during cooking (Heron et al. 2016), it began to be used to identify broomcorn millet processing in pottery (Ganzarolli et al. 2018; Heron et al. 2016). Because the compound is not particularly abundant in the grain, identification of miliacin is most easily performed using SIM mode (Heron et al. 2016). This is a rare example of a staple grain that contains a known biomarker.

Hopane triterpenoids were mentioned in passing in the Deepwater Horizon case study described in chapter 3. Unlike many triterpenoids, these compounds derive from a wide range of resources. They are a major constituent in dammar gum (Langenheim 2003) but are also commonly formed by bacteria. Hopanoids therefore are often used as a biomarker for oil and hydrocarbon presence in a sediment or residue; in fact, hydrocarbons can often be sourced to a specific oil field through the ratios of hopanes and steranes in a given hydrocarbon (Meyer et al. 2018). Interestingly, some hopanoids are also used as a biomarker for pulque production, as the bacteria primarily used to ferment agave into pulque produces unusually abundant hopanoids

(Correa-Ascencio et al. 2014). Hopanes, therefore, can be used in a variety of different residue interpretations, depending on the precise hopanoid identified and the archaeological and cultural context. Hopanoids typical of crude oil or bitumen might reflect contamination in a residue from a Louisiana site exposed to an oil spill. When found in a vessel with an unusual visible residue, it might reflect bitumen processing or pot sealing. Hopanoids typical of dammar gum probably reflect resin processing or the processing of an unusual resinous plant. And particular bacteriohopanoids may be used to indicate alcohol production in some contexts.

Given the wide range of triterpenoid compounds, it seems possible that more triterpenoid biomarkers for specific classes and resources are present in the natural world. Establishing a biomarker, however, is difficult. First the compound must be identified as present in a specific resource. Then its functional absence in other resources must also be established. Further, it must also be established, or at least reasonable, that the compound will be able to survive in residues during archaeological deposition. It seems logical that future in-depth analysis of resources of interest will produce at least some triterpenoid biomarkers as useful as betulin, miliacin, and the various hopanoids.

PHENOLIC RESINS

Terpenoids are not the main ingredient of all resins. For example, the American sweetgum (*Liquidambar styraciflua*) produces a liquid resin from wounds in the trunk that, when dried, appears almost indistinguishable from terpenoid resins. It hardens with age, has a sweet scent (hence the name of the tree), and appears to have cultural uses almost indistinguishable from the uses of terpenoid resins, including incense, wound care, and chewing gum (Lingbeck et al. 2015). Chemically, however, the sweetgum resin is quite different from a terpenoid resin, as it is composed primarily of cinnamic acid and its derivatives (Langenheim 2003; Lingbeck et al. 2015). Cinnamic acid, shown in Figure 5.8, is a phenolic acid and not a terpenoid. As such, the compound itself is not as stable in residues as terpenoids. As the resin hardens, the cinnamates link together into larger structures, however, producing a natural polymer. During residue extraction, they will tend to break apart, producing cinnamic acid and other cinnamates, which can be used as an indicator that an unusual plant resin is present in the residue. Unfortunately, cinnamic acid and its derivatives are found in a wide range of plants, not just resins, including such sources as sunflowers (*Helianthus anuus*) and peppers (*Capsicum anuum*) (US Department of Agriculture, Agricultural Research Service 1992–2016). Still, abundant presence of cinnamic acid in a residue may indicate the presence of a phenolic resin.

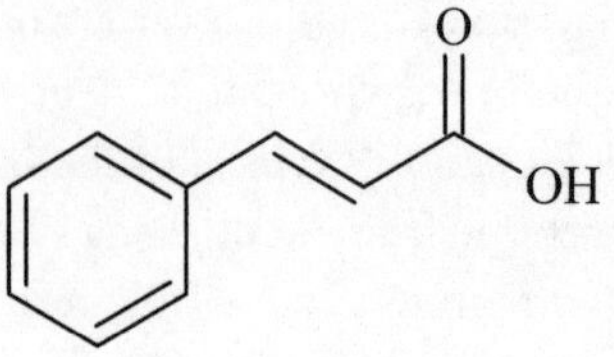

Figure 5.8. Molecular structure of cinnamic acid, the most important molecular structure composing sweetgum resin. This is a phenolic acid, rather than a terpenoid, and is chemically distinct from many other resins. (Eleanora A. Reber)

In addition, some phenolic resins may have psychoactive qualities, such as the resins found in marijuana (*Cannabis* spp.). These are a mixture of phenolics and light terpenoids (Langenheim 2003). The terpenoids are unlikely to survive in an archaeological residue, but the phenolics, which include the psychoactive cannabinoid compounds, would be likely to survive in archaeological residues with some oxidation and hydrolysis and have been detected using untargeted analysis (Ren et al. 2019).

In general, most analysis of resins in residues has focused on terpenoid resins. A variety of other resins and lacquers derived from phenolic compounds (sometimes called benzoins) occur in a wide range of plants, however, and should be possible to detect in residues (Langeneheim 2003). The archaeologist and analyst should communicate with each other, and possibly with an ethnobotanist, about which resins might be found or used in the site area. This will help considerably with interpreting the source of any unusual phenolic compounds found in the residues.

NUTS AND TERPENOIDS

In much of the midwestern and eastern United States, many different types of nuts were important in indigenous American diets. These high-fat, high-protein foods were eaten from the shell as well as processed into soup, breads, and energy drinks (Fritz et al. 2001; Swanton 1979), probably making up a large amount of the diet wherever nuts were abundant. Given the known dietary importance of nuts, it would be exciting to be able to identify them in pottery residues.

Although nuts are high in fats and proteins, and therefore would be expected to contribute to residues whenever they were processed in pots, they do not seem to be particularly well endowed with biomarkers. Lipid extracts of nuts, as shown in Figure 5.9, are primarily composed of triacylglycerols (TAGs), which are themselves made up of common fatty acids. Pure nut residues clearly look like plant-based foods, but they cannot be distinguished from other plant-based lipids.

The most likely avenue to identify nuts in absorbed residues derives from the identification of triterpenoid compounds present in nutshells. Hickory

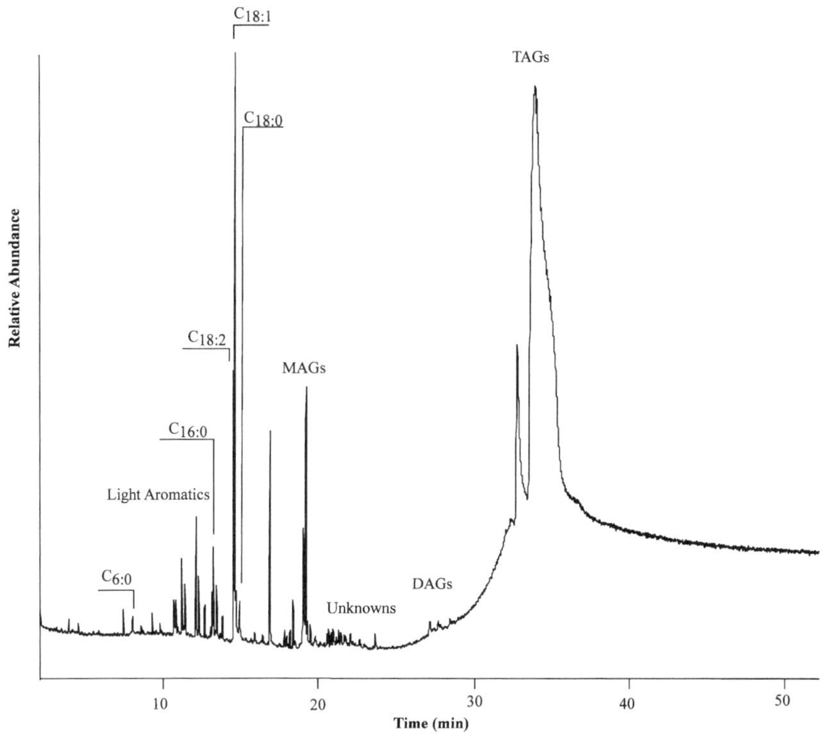

Figure 5.9. Total lipid extract of lipids extracted from a pecan nut, *Carya illinoinensis*, using chloroform/methanol extraction. TAGs (triacylglycerols) dominate the residue, which is composed almost entirely of acyl lipids typical of plant foods. Unusual compounds that might act as biomarkers are not present. MAGs = monoacylglycerols; DAGs = diacylglycerols. (Eleanora A. Reber)

nuts are a particularly valuable food source in much of the southeastern United States, but the nuts may be difficult to remove from their thick, deeply grooved shells. At least some indigenous American people solved the dilemma of hickory production by pounding the hickory nuts, shells and all, in mortars and pestles. The large pieces of shell were then removed, and the remaining mixture of nutmeats and shells was pounded into a paste, which was formed into a ball (Fritz et al. 2001). This ball could then be added to water to produce a soup, called ku-nu-che (Fritz et al. 2001). The processing of both the nuts and the nutshells together might allow the extraction of triterpenoid compounds from the shells. Further research on triterpenoids in hickory nut shells might provide an indicator for the presence of ku-nu-che processing in a pot. Unfortunately, analysis of nuts without shells has not yet identified any unique terpenoids or other biomarkers.

THE UBIQUITY OF RESINS

After some years of experience in analyzing residues, it seems clear that diterpenoid and triterpenoid compounds are more common in pottery residues than I originally expected. The reason for this seems to derive from two major factors. One is technical—because the compounds are stable and do not dissolve in cold water, it is probably almost impossible to get them out of residues once they have absorbed within the pores of a pot or charcoal encrustation. Unlike unsaturated fatty acids, these terpenoids will not easily disappear from a residue. If they do oxidize or hydrolyze, it is into another, stabler diterpenoid or triterpenoid. The other reason probably derives from the usefulness of plant resins to people. Botanically, the purpose for these resins is often as an insect repellent or antiseptic—they are meant to protect their host plants from a variety of invaders, whether they be insects or microbes. As people are also interested in repelling insects and killing microbes, it makes sense that plant resins were often used in a variety of ways. The texture, smell, and taste of resins and resinous plants also tend to mark them as useful. Consider the many uses for resinous plants discussed just in this short chapter: vitamin C supplement, sealant, glue, incense, cold medicine, occasional psychoactive, and flavoring. Although modern people do not use resinous plants extensively, clearly ancient people did use them, and in a wide variety of ways.

Because resins from different plants and trees can look chemically similar in a residue, communication between the residue analyst and both the archaeologist and an ethnobotanist is particularly helpful in interpreting these resins. Often ethnobotany provides crucial information that allows resin-based compounds in residues to be interpreted more specifically and usefully than could otherwise be the case.

6

COMPOUND-SPECIFIC ISOTOPIC ANALYSIS

Detecting Maize and Isotopically Unique Resources in Residues

Isotopic analysis provides another approach to residue analysis when biomarkers are not present in a residue. In the absence of biomarkers, this type of analysis can detect the presence of isotopically unique grains, marine fish, ruminant fat, ruminant milk, and dairy products. For this type of analysis to work, however, the resources detected must be isotopically different from other resources in the environment.

Isotope analysis can be applied to residue analysis most sensitively using compound-specific isotope analysis (CSIA), a relatively recent instrumental development that allows compound-level identification of isotopically unique compounds. Understanding and interpreting the results of this analysis can be complex, as it requires an understanding of isotopic chemistry, photosynthesis, food webs, and the mixing of compounds within organic residues. This chapter gives background on all of these areas and also shows how they interact to produce remarkably powerful, sensitive detection of isotopically unique resources in residues. This chapter also includes a more detailed discussion of how lipids are absorbed within the ceramic matrix, which is important in discussing the concept of how lipids from lipid-rich resources can overwhelm the isotopic signature from resources with fewer lipids.

BRIEF BACKGROUND ON ISOTOPES AND RESIDUES

Compound-specific isotope analysis is created by attaching an isotope ratio mass spectrometer (IRMS) to the end of a GC or HPLC instead of the standard mass spectrometer. The IRMS is a highly accurate, tuned mass spectrometer that measures individual stable isotopes of different elements in a molecule instead of larger molecular structures and fragments. The IRMS can be tuned to detect isotopes in many different elements that are useful to multiple fields of study, including carbon, oxygen, hydrogen, and nitrogen. Stable isotope analysis is often used archaeologically in the analysis of human skeletal remains and can provide unique information on a variety of topics. For general reference, Table 6.1 describes some of the most commonly used elements and applications for different stable isotopes in archaeology.

Most lipids are made of carbon, oxygen, and hydrogen, so carbon is the most useful type of stable isotope analysis to apply to absorbed residues. Although in theory, oxygen and hydrogen analysis could be performed on lipids in pottery residues, oxygen and hydrogen isotopes generally provide information on rainfall patterns. Heavier isotopes fall from the clouds as rain sooner, so that lighter ^{18}O and ^{2}H ratios tend to occur in areas farther away from the dominant source of clouds and rainfall, usually a large body of water. In most circumstances, foodstuffs aren't traded such long distances that differences in rainfall pattern become apparent, and there are easier artifacts to sample for ^{18}O and ^{2}H ratios if the archaeologist is interested in rainfall pattern shifts. In addition, sampling residues for ^{2}H ratios is complicated because there is a possibility that hydrogen in residues might exchange with hydrogen from groundwater or other depositional hydrogen sources (Schimmelman et al. 2006).

Nitrogen isotope analysis is not possible on lipids because most lipids don't contain nitrogen. Although nitrogen is present in alkaloids, these compounds usually derive from plants, and so information on trophic shifts and the length of the food chain is not present. If present in compounds, nitrogen isotopes could potentially determine field fertilization, based on enriched $\delta^{15}N$ values caused by preferential excretion of the heavier isotope of nitrogen in manure (Choi et al. 2003; Simpson et al. 1999; Szpak et al. 2012). The presence of very dry growing environments could potentially be detected through compound-specific nitrogen isotope analysis of alkaloids or other nitrogen-containing compounds. Rainfall seems to preferentially wash larger amounts of ^{15}N out of soil than the lighter ^{14}N, meaning that drier areas have higher $\delta^{15}N$ values in the soil, which is then passed on to the vegetation (Ambrose 1991; Austin and Vitousek 1998). This technique has not yet been applied, however, to the best of my knowledge. Nitrogen isotope analysis may be applied to visible residues, as these encrustations often include proteins and other nitrogen-containing compounds.

The information derived from CSIA of lipids is largely parallel to that given by stable carbon isotope analysis of bone collagen and apatite, which is widely used in archaeology. To understand the application of carbon stable isotope analysis to residues, it helps to review how this analysis works overall in archaeological interpretation and the ways it interlinks with issues of photosynthesis and food chains.

Table 6.1. Some Common Stable Isotopes and Their Uses in Archaeology

Isotope	Material	Information provided
Carbon (^{13}C, ^{12}C)	Bone collagen, bone apatite, lipids, seeds, any carbon-containing artifact	Dominant photosynthetic pathway of plants, intrusion of plants with a different photosynthetic pathway into the ecosystem, presence of marine elements
Oxygen (^{18}O, ^{16}O)	Bone apatite, bone collagen, hair; any oxygen-containing artifact	Rainfall patterns, suggesting either intrusion from areas with other rainfall regimes or changes in weather patterns
Nitrogen (^{15}N, ^{14}N)	Bone collagen, hair, proteins, amino acids; any nitrogen-containing artifact	Trophic shift, suggesting how high an individual ate on the food chain, presence of marine elements, presence of leguminous plants, presence of very dry environments
Hydrogen (^{2}H, ^{1}H)	Bone collagen, bone apatite, lipids, hair; any hydrogen-containing artifact	Similar to oxygen-rainfall patterns, suggesting either intrusion from areas with other rainfall regimes or changes in weather patterns

BRIEF BACKGROUND ON CARBON STABLE ISOTOPE ANALYSIS IN ARCHAEOLOGY

Carbon stable isotope ratios (sometimes written $\delta^{13}C$) were originally applied to archaeological remains in 1977, when Vogel and van der Merwe published their seminal study on the identification of maize consumption in prehistoric remains from New York State (Vogel and van der Merwe 1977). The analysis works because of the photosynthetic origin of the carbon making up the world's food webs and the tissues of all living creatures.

Plants use photosynthesis to incorporate carbon from atmospheric carbon dioxide into their tissues, producing lipids, carbohydrates, and every other carbon-based structure of the plant out of (literally) thin air. Because of wind and atmospheric mixing, the ratio of the stable carbon isotopes ^{13}C to ^{12}C ($\delta^{13}C$) is mostly constant in the global atmosphere. These isotopes are stable, or not radioactive, so this ratio stays the same. That means that if every terrestrial plant in the world had the same method of photosynthesis, the $\delta^{13}C$ of all contemporaneous plant tissues would be broadly similar, taking into account variation caused by rainfall and the amount of tree canopy present, which may slightly shift the local $\delta^{13}C$ ratio (Hare et al. 2018).

As other creatures ingest plants, the body and cells of each plant eater transform the plant tissues into the tissues and compounds most useful to that individual. Atoms of ^{13}C are slightly larger and heavier and therefore react slightly differently from atoms of ^{12}C in the various biochemical processes that take place in the body of a plant-eating animal. In some cases, the lighter ^{12}C tends to be exhaled preferentially in the form of carbon dioxide, resulting in body tissues that are slightly more enriched in the heavier ^{13}C (DeNiro and Epstein 1978). This process is called a trophic shift, and it permits some detection of how high on the food chain an individual is eating through carbon stable isotope analysis. Trophic shifts vary, however, based on the type of biochemical processes taking place in the body and the way each molecular type is transformed. Lipids are often incorporated directly from a food into the body of the consumer with minimal transformation. That means that lipids have either a very small or no trophic shift, or even a shift toward less ^{13}C rather than more—the reverse of the type of shift expected in, for example, bone tissue (DeNiro and Epstein 1978; van der Merwe 1982). Variable $\delta^{13}C$ values in plants, however, are passed up through the food chain with varying degrees of very small trophic shift, depending on the body tissue being measured.

NOTE: WHY ARE ALL THE NUMBERS NEGATIVE?

The number reported as $\delta^{13}C$ is a ratio compared with a known standard. For historical reasons, this was a piece of Pee Dee Belemnite from South Carolina. The original standard is now long gone, but by tradition carbon isotope values are always given with reference to this standard, which is called PDB. The belemnite was a marine rock formation and had an unusually high amount of ^{13}C compared with ^{12}C. As a result, most samples measured for stable carbon isotopes have less ^{13}C than the standard and are given as a negative number. The variation in these isotopes is also very small, generally in the parts per thousand range. As a result, stable carbon isotope results are usually given in the form $\delta^{13}C$ X.X‰PDB, in which X.X is a negative number. The permil notation (‰) saves us from writing all results in the form 0.00XX. As results are usually negative compared with the standard, numbers that are less negative, for example –4‰, have more ^{13}C in them than those that are more negative, such as –27‰. Samples with more ^{13}C than an expected value are sometimes referred to as "enriched" relative to that value. Samples with less ^{13}C than typical are sometimes referred to as "depleted" relative to an expected value or another resource.

As a result, the primary cause of differences in stable carbon isotopes is the photosynthesis used by the plants when they fix atmospheric carbon into their tissues. As mentioned, if all plants used the same photosynthetic methods, there would be little variation in terrestrial stable carbon isotopes. There are three different photosynthetic pathways, however. C_3 photosynthesis was the first type to evolve. It fixes ^{12}C more efficiently than ^{13}C, and so C_3 plants show a sharply more negative (depleted) $\delta^{13}C$ value than the atmosphere, with most tissue $\delta^{13}C$ values between –21 and –36‰. C_4 photosynthesis independently evolved many times around the world to optimize photosynthetic efficiency especially in sunny areas. The pathway works by concentrating carbon dioxide within the cells near the photosynthetic sites. The result is a more efficient and faster photosynthetic process in bright sunlight, and less depletion of ^{13}C in the tissues than in C_3 photosynthesis. Tissue $\delta^{13}C$ values are usually between –10 to –19‰. Crassulacean acid metabolism (CAM) is a photosynthetic pathway adapted for photosynthesis in arid environments and is found almost exclusively in succulent plants. The $\delta^{13}C$ values of CAM plants may be intermediate or overlap between those of C_3 and C_4 plants (Cernusak et al. 2013). The $\delta^{13}C$ values are passed up through the food chain, so the presence of different photosynthetic pathways can be determined in the tissues of all creatures (including humans) in a relevant food web—the tissues of a carnivore will reflect the dominant photosynthetic pathway of the plants eaten by their prey at the base of the food chain.

The application of this information is well known in archaeology, particularly with respect to studying human bone collagen and apatite to detect the advent of an important food crop with a different photosynthetic pathway than that of the dominant ecosystem in an area. The original application of the technique in archaeology was detecting the introduction of the C_4 grass maize into the predominantly C_3 environment of northeastern and midwestern North America (van der Merwe and Vogel 1978; Vogel and van der Merwe 1977). The technique can be used, however, to note any shift in the photosynthetic pathway of a major part of people's diet in an area—the advent of rice (C_3) into an area previously using millet (C_4), for example. The technique is not useful when a domesticated plant has the same photosynthetic pathway as the dominant foraged foods in an area. For example, it will not show the adoption of millet (C_4) in South Africa, as the ecosystem is predominantly C_4 in that area, as are most of the foraged, nondomesticated plants (van der Merwe 1982). Similarly, carbon stable isotopes do not change in southeastern North America with the adoption of the Eastern Horticultural complex, a suite of C_3 plants domesticated in the southeastern United States; these domesticated plants are C_3, like their wild precursors.

Stable isotope analysis originally focused on the human bone tissues collagen and apatite as well as tooth enamel. As the photosynthetic pathway affects all parts of a plant, however, it also applies to residues, both in bulk and individually to each compound in a residue.

APPLYING STABLE ISOTOPES TO BULK RESIDUES

The interpretation of carbon stable isotope analysis in residues is different from that of bone tissues. For one thing, isotopes in residues give information on resources processed in pots, and not necessarily resources directly eaten. As in all residues, resources not processed in pots will also not be identified through pottery residues. Bone tissue results give a kind of average of the entire diet (or protein intake, in the case of collagen) of an individual over the past time period in which the bone remodeled. Residue isotope results give an average of the items processed in the pot, which occurred over a variable period of time that can't easily be determined. In addition, the trophic shift of lipids is different from that of collagen and apatite, so that the baseline values are different. Results with the same meaning are actually numerically different.

Unseparated, or bulk, residues underwent stable isotope analysis before compound-specific analysis for instrumental reasons. The original mass spectrometers could not easily be linked to GCs or HPLCs, and so compound-specific isotope analysis of residues was not technically possible until the late 1980s. Instead, samples were turned to carbon dioxide through combustion of an entire sample, and the carbon dioxide was introduced into the mass spectrometer in a single burst. The result was a $\delta^{13}C$ value for an entire residue. Visible residues on the surface of pots can be analyzed in bulk by simply combusting scrapings of the residue. As visible residues often include protein as well as lipids, this had the benefit of allowing $\delta^{15}N$ analysis as well as $\delta^{13}C$, which allows meat, fish, and plant contributions to the residue to be more easily determined (see Table 6.1). Because of this, it is often fruitful to use bulk isotope analysis on visible residues.

Absorbed residues could also undergo bulk isotope analysis during this period, but it was technically difficult and tended to give unsatisfying results. The usual method for bulk residue collection of absorbed residues involved introducing small amounts of the residue into tiny tin foil autocombust cups and then carefully evaporating the solvent to produce pure residue deposited on the interior of the cup. The cup and its residue were then combusted to produce carbon dioxide for analysis. From a practical standpoint, if the evaporation was not done carefully enough, the tin foil cup was blown out of its container and went flying across the fume hood, potentially contaminating its contents and requiring the analyst to resample.

Furthermore, multiple factors complicated interpretation of bulk results for both absorbed and visible residues. Contaminants such as plasticizers, if present in the residues, would necessarily be present in the bulk isotope results and skew them. In addition, the bulk samples analyzed the isotope values of many different compound types simultaneously, and different types of compounds have different trophic shifts. These trophic shifts depend on the synthesis of each compound in a plant and then on any further remodeling or biochemical reactions that take place as the compounds travel through the food web.

A specific complication with bulk analysis of residues lies in varying food contributions to residues. Experimental analysis demonstrated that bulk isotope analysis did not seem to detect C_4 grain, especially maize, in amounts reflecting its known presence in experimentally produced residues (Hart et al. 2007). An isotopic approach was needed that allowed a finer-grained approach to residue analysis and to multicomponent samples in general. Compound-specific isotope analysis, when it was developed, allowed this type of in-depth approach to residues and stable isotopes. In addition to all its other benefits, plasticizers and other contaminants can simply be discounted from the overall data, solving many problems.

A Note on Lipids and the Carbon Cycle: Some Rules of Thumb

As we investigate CSIA of lipids more thoroughly, some of the details can seem complex. Detecting isotopically unique plants requires an understanding of photosynthesis. When we discuss marine and ruminant lipids, below, we will delve into the complexities of the marine carbon cycle, the dietary action of microbes in the rumen, and fatty acid synthesis in the mammary glands of cattle. As CSIA becomes more established, it seems likely that even more interesting results can be obtained through careful investigation of the carbon cycle. To help simplify these analyses, here are some basic rules for dealing with lipids and the carbon cycle:

(1) You are what you eat—the first law of stable isotopes! The basis of any food web is the carbon source sustaining it. Marine carbon is a different source than atmospheric carbon, and so the entire marine food web is isotopically different from the terrestrial one. When in doubt, determine the original source of the carbon (atmospheric, marine, etc.) and the type of photosynthesis present.

(2) All animal bodies are efficient and prefer to use compounds exactly as they appear in their foods, with no processing. Every herbivore and carnivore will preferentially use fatty acids and other lipids as they appear in their diet.

(3) If there are more lipids in the herbivore or carnivore than present in their diet, the lipids must come from somewhere else. The body will have to synthesize the new fatty acids from other parts of the diet, probably carbohydrates. This will lead to an isotopic enrichment of carbon in the lipids of the animal in question because carbohydrates are more enriched than lipids. The synthesis will also likely cause some trophic shift, but only in those lipids that are synthesized.

(4) With carbon, as long as the lipids in the food match with the lipids needed by the animal eating it, there will not be a shift in the isotopic values of the lipids, because of #2 above. This is only true for carbon, not for nitrogen or other isotopes, which follow their own individual cycles with their own rules of thumb.

COMPOUND-SPECIFIC STABLE ISOTOPE ANALYSIS AND HOW IT WORKS

In CSIA, as described earlier, the residue is run through a GC column, the same as in GC/MS. Instead of feeding directly into a mass spectrometer, however, each peak coming off the GC column is routed into a combustion chamber, where it is burned into carbon dioxide. The carbon dioxide pulse then continues into the stable isotope mass spectrometer, which continuously monitors the isotopes of interest—in this case ^{13}C and ^{12}C. The results are plotted similarly to a GC/MS, except the MS of each GC peak now identifies the $\delta^{13}C$ value of each separate pulse of carbon dioxide rather than the molecular fragmentation patterns. This allows separate $\delta^{13}C$ measurements for each peak in a residue, as long as the peaks separate adequately. This type of instrumental analysis is often called GC-isotope ratio mass spectrometry, or GC-IRMS. Since the MS will only measure $\delta^{13}C$ values, the compounds analyzed on the GC-IRMS must be identified by comparison with parallel GC/MS runs. The technique was originally used as an application to identify the illegal adulteration of olive oil with corn oil (Woodbury et al. 1987; 1995) but was soon applied to archaeological pottery residues (Dudd et al. 1999; Raven et al. 1997, and others). Once this application began, it was clear that CSIA analysis of residues allowed a wide range of interpretations. It permitted the detection not only of the adoption of maize and other C_4 plants in C_3 environments (and vice versa) but also the presence of marine resources, ruminant fats, and dairy products in residues (Copley et al. 2005; Craig et al. 2007; Mukherjee et al. 2007, and many others). An IRMS laboratory is shown in Figure 6.1; this facility can perform many different stable isotope analyses, including either bulk or compound-specific.

Figure 6.1 Isotope Ratio Mass Spectrometry facility at UNCW. This sequence of instruments permits compound-specific isotope and bulk stable isotope analysis of pottery residues as well as a variety of samples from other disciplines. Depending on the application, an auto-combust, direct injector, or GC is linked to the combustion chamber, which then feeds the combusted sample into the isotope ratio mass spectrometer. (Courtesy Chad Lane)

A Note on Calibration, Derivatization, and GC-IRMS

GC-IRMS essentially weighs carbon atoms and is very sensitive. As a result, the analyst must address a number of technical issues to ensure that the data are accurate. First, the $\delta^{13}C$ value for each peak is measured in comparison to internal spikes (artificially added peaks) of a known carbon dioxide reference gas. It's crucial that the $\delta^{13}C$ value of the reference gas is calculated accurately. Furthermore, a number of factors can cause drift and uncertainty in measuring these very small weights. To check for problems, calibration runs of compounds with known $\delta^{13}C$ values are important and should take place regularly throughout an instrument sequence. When we run samples on GC-IRMS, one calibration compound is analyzed for every four archaeological analyses. If the results start drifting away from the known values of the calibration compounds, analysis must be stopped and the instrument readjusted (recalibrated, a *seed oxidation initiated to clean the combustion chamber, the combustion chamber swapped out, etc.) in order to get accurate results. In addition, we run a blank analysis (injection of clean hexane only) every four archaeological samples and after the calibration runs to make sure that there is no contamination on the column. A small amount of a compound remaining on a column from a previous run can contaminate later analyses.*

All potential sources of nonarchaeological carbon must also be controlled for and added into the results. For example, fatty acids are often derivatized to methyl esters, which include one modern carbon atom added to each archaeological fatty acid. The $\delta^{13}C$ value of the source of this carbon atom is measured so that it can be discounted from the value measured by the mass spectrometer. To ensure that the instrument is measuring consistently, every sample is run at least in duplicate. If peaks are small, or results seem skewed, it may be run in triplicate or even more, to make sure that the results are repeatable. As a result, every CSIA result for a residue peak contains multiple sources of potential uncertainty: detection limits on the reference gas, variation on the calibration runs, variation in detecting the derivatizing carbons, and variation between duplicate runs of the same sample. After all this, how are you supposed to know whether the results are accurate within a reasonable uncertainty? There are various ways of measuring all these uncertainties. If one is calculating each source of uncertainty separately, no one of these areas should be more than 0.5 standard deviation (SD). In our lab, I calculate a pooled SD that includes all of these potential sources of uncertainty into a single number (Polissar and d'Andrea 2014). A pooled SD of 0.5 is optimal, but 0.7 is acceptable, especially when running samples with high variation between peak sizes.

This is because it's important to get enough carbon dioxide into the mass spectrometer with every peak so that an accurate measurement can be made, but not so much that it overwhelms the instrument. Too much or too little carbon dioxide will cause a completely inaccurate result. Fortunately, these inaccurate results are clearly detectable by the shape and size of the peaks appearing on the GC trace. When injecting residue samples into a GC-IRMS, residues containing large peaks, such as fatty acids, must be diluted with clean solvent so that the fatty acids don't overwhelm the spectrometer. Residues containing very small peaks, such as sterols or alkanols, should be concentrated as much as is practicable and injected in larger amounts, so that the peaks will be large enough for accurate detection. Residues containing both types of compound, such as residues extracted with acidic methanol, must often be run twice (in duplicate), once at the fatty acid concentration and another at the neutral compound concentration. The neutral concentration run will vent the fatty acids before combustion so as to prevent unnecessary wear and tear on the combustion chamber and mass spectrometer.

When selecting a laboratory for CSIA analysis of pottery residues, it's best to choose one with some experience in archaeological residue analysis, as the uncertainties are wider and more complex than with many other types of samples, and the types of samples run are more complex than most nonarchaeological samples. Your analyst should be able to discuss how they control for uncertainties and their criteria for accepting or rejecting $\delta^{13}C$ values of compounds in a sample.

Case Study: Direct Detection of Maize in Absorbed Pottery Residues

Having worked through all the technical issues, we can now discuss the more interesting elements of interpretation. For example, how does CSIA allow a clearer picture of maize adoption in midwestern North America? As mentioned, bulk isotope analysis of a residue includes all compounds in a residue, including contaminants and a range of archaeological compounds with slightly different trophic shifts. CSIA allows each compound to be analyzed and compared separately, breaking apart different elements of the residue isotopically.

Figure 6.2 shows a comparison of carbon isotope values for the most common fatty acids of a series of known modern resources, excluding marine and anadromous fish, and a series of archaeological samples that I've analyzed since 2000. The circle indicating a probable C_4 or CAM contribution was drawn by determining the most enriched C_3 plants and freshwater fish values and continuing to the most enriched C_4 plant lipid values. As can be seen from Figure 6.2, a C_4 or CAM contribution is present in some of these residues, especially those analyzed in 2000–2001 in a project focusing on maize adoption across the Mississippi Valley. Despite the presence of a C_4 or CAM contribution to some of the residues shown in Figure 6.2, however, we can't definitely say that maize was the source of this contribution to the fatty acids.

The values in Figure 6.2 indicate the presence of a C_4 or CAM plant in the food web, not necessarily of the plant itself in the residue. Fatty acids from animal products derive their $\delta^{13}C$ values from the plants at the base of the food chain with relatively little trophic shift. C_4 or CAM values in fatty acids could therefore come from animals, such as deer, that ate maize during the months prior to their death. Furthermore, the isotope values of maize could be mimicked by fatty acids from a small number of native C_4 and/or CAM

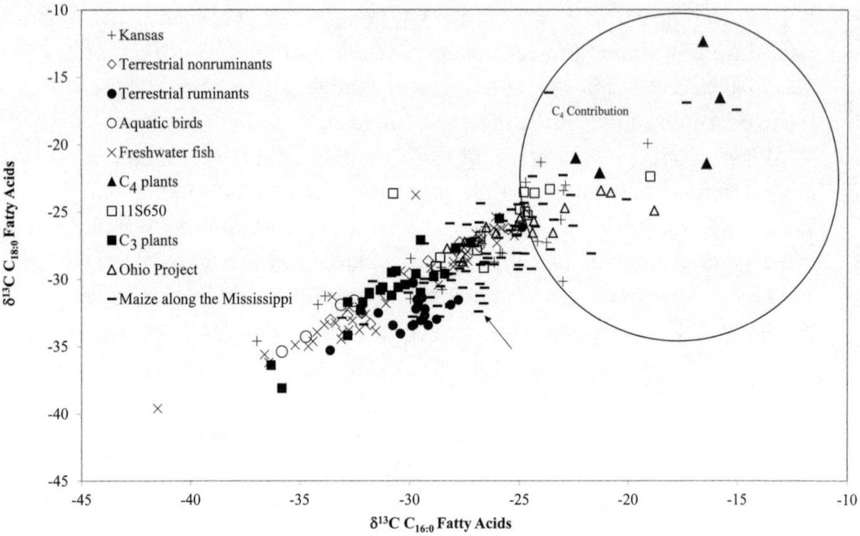

Figure 6.2. δ^{13}C values for $C_{16:0}$ and $C_{18:0}$ fatty acids from all archaeological samples analyzed from 2000 to 2019, compared with those from known modern resources. These have been analyzed in our lab and also published by many other scientists (Ballentine et al. 1996; Choy et al. 2016; Craig et al. 2011, 2012, 2013; Fernandes et al. 2018; Spangenberg 2016; Spangenberg and Ogrinc 2001; Steele et al. 2010; Taché and Craig 2015; Woodbury et al. 1998). Fatty acids with a C_4 contribution are in the circled area. The arrow points to HRL 208, discussed in more detail in the text. (Eleanora A. Reber)

plants in the American Midwest. The most well-known of these are amaranth (*Amaranthus* spp.), prickly pear cactus (*Opuntia* sp.), common purslane (*Portulaca oleracea*), and native North American panicgrass or switch grass (*Panicum* spp.). None of these plants had the economic or dietary importance of maize, but they are known to have been occasionally grown and consumed in the largely C_3 environment of midwestern and southeastern North America and could contribute to pottery residues.

To be sure that the C_4 contribution in a residue comes from maize and not from an animal eating maize or from nonmaize C_4 or CAM plants, some form of isotopic biomarker is necessary. That would be a compound that is present in plants and not animals and that is (ideally) present and abundant in maize but not present or abundant in purslane or prickly pear. When the original maize-detection project began in 2000, I had planned to do compound-specific analysis on sitosterol, a common plant sterol. Although it is present in nearly all plants, including purslane, amaranth and prickly pear, it would not be present in nonplant resources. Sitosterol with C_4 values

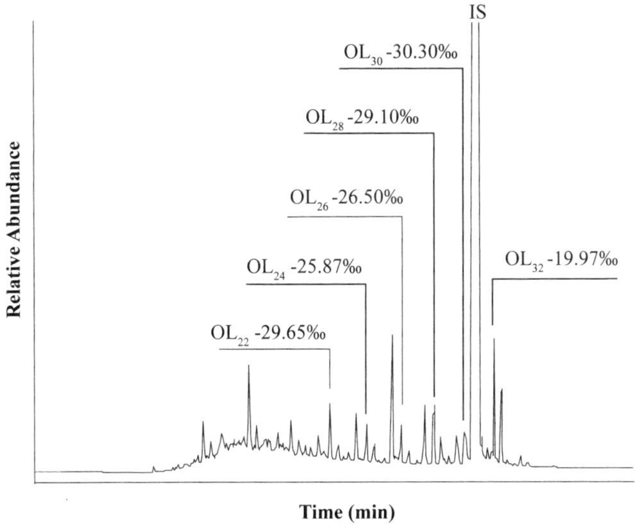

Figure 6.3. Neutral fraction from HRL 208, showing the difference in stable isotope values for different alkanols in the same fraction. Note especially the difference between *n*-dotriacontanol (OL_{32}), *n*-tetracosanol (OL_{24}), and the other alkanols in the fraction. (Eleanora A. Reber)

would permit the identification of C_4 plants in a residue, and we could then rely on paleoethnobotanic records to determine whether maize was the most likely plant to produce the C_4 contribution.

As I analyzed the data, however, I noticed several residues with intriguing results. Most noticeable was a residue (HRL 208) extracted from a black-slipped sherd from the Osceola site, a Coles Creek site in Louisiana, dating ca. AD 700–1200. Maize was present at the site in relatively low densities, although there was one feature containing a large amount of the crop (Fritz and Kidder 1993). The fatty acids from the sherd had C_3 values, as shown by the arrow in Figure 6.2. The compound-specific results of the neutral fraction from the residue, however, did not have completely C_3 values. The neutral fraction, shown in Figure 6.3, didn't contain enough sitosterol or plant sterols for compound-specific analysis but did contain an abundant sequence of alkanols. The alkanols with 22 to 30 carbon in the even chains had C_3 values, like the fatty acids (alkanols in natural products generally are even-chain, as mentioned in Table 1.1). The $\delta^{13}C$ value of the 32-carbon alkanol, *n*-dotriacontanol (OL_{32}), however, was typical of C_4 plants, especially compared with the other alkanols and fatty acids in the residue.

Soon, I began to observe that other neutral fractions contained *n*-dotriacontanol, and that in some cases this compound was more enriched than other compounds in the residue. In other cases, residues with C_4 derived fatty acids also had C_4 derived *n*-dotriacontanol in their neutral fractions. Every residue with a C_4 contribution to fatty acids that contained measurable amounts of OL_{32} also had *n*-dotriacontanol with a C_4 contribution. Some residues also contained fatty acids and *n*-dotriacontanol that all originated in C_3 resources.

Intrigued, I went back to the literature on maize lipids, which is extensive. Although at the time no one had performed basic research on the compound-specific isotope values of the alkanols in maize oil, there was analysis of the composition of these lipids. When focusing on the alkanol portion of the lipids, it was clear that maize had an unusually large amount of OL_{32}, especially compared with the usually common OL_{28}, which is relatively less abundant in maize (Bianchi et al. 1984). Maize alkanols also contain OL_{24}—and when looking at Figure 6.3, it is clear that OL_{24} is the next most enriched alkanol in the sequence after OL_{32}. Usefully for residue analysts, *n*-dotriacontanol is not normally a common alkanol, although it is often abundant in panicoid grasses (Tulloch 1981). This category includes maize but also a large subfamily of more than 3,500 other grasses, which includes panicgrass (*Panicum* spp.) and many other grasses, some of which (millet, sugar cane, sorghum) are economically important. *N*-dotriacontanol is also presumably abundant in some nonpanicoid grasses. However, the fact that the compound is more abundant than usual in maize, and relatively rare in nonpanicoid grasses, is useful. It not present in purslane (Tulloch 1974), although it is present in at least some species of *Opuntia* (Wilkinson and Mayeux 1990). In addition, the compound is most common in plant waxes, although it is occasionally present in insect waxes, notably lac (Chibnall et al. 1934). As a result, *n*-dotriacontanol acts as an isotopic biomarker for the presence of C_4 plants, possibly maize, in C_3 environments. Enriched OL_{32} can act as a clear indicator that a C_4 plant contributed to a residue. It also indicates that the C_4 plant was not purslane or prickly pear and may have been a panicoid grass.

Sample HRL 208 was particularly instructive, as its fatty acids did not show a C_4 or CAM contribution, but its *n*-dotriacontanol did. How could this sample show such enriched OL_{32} and such depleted fatty acids and other alkanols? To answer this question, we must further investigate the ways in which residues form.

Masking

As described, absorbed residues are largely made up of lipids. Naturally, each different resource processed in a pot can only contribute the lipids that it contains. That means that common compounds, such as fatty acids, are contributed by almost every different resource processed in the pot, while only some of the resources contribute rarer compounds like OL_{32}. If an isotopically unique resource, like maize, is only one of the resources processed in a pot, its contribution to common compounds will be mixed with, or even completely swamped by, the contributions from other resources. Its contribution to rarer compounds will be less mixed or in rare cases derive only from maize. This is what happened in HRL 208. This residue includes compounds from a range of resources. In this case, maize appears to have been the only C_4 resource processed in the pot, producing the highly enriched OL_{32}. In terms of its lipid contributions other than OL_{32}, however, maize seems to have been a minor presence in the pot. The depleted fatty acids contributed by all the nonmaize resources in the residue overwhelmed the enriched fatty acids contributed by the maize. The same thing happened to the alkanols, except for OL_{32} and OL_{24}. The result was a residue-containing lipids typical of C_3 photosynthesis . . . except for the *n*-dotriacontanol.

A similar process happens with all isotopically mixed residues, although it's not usually as obvious as in HRL 208. Different resources contribute different compounds according both to the amount of lipid present in a resource and the abundance of each compound in that resource. It therefore is important to note that depleted fatty acid $\delta^{13}C$ values don't actually mean the residue is composed completely of C_3 lipids. Instead, it means that only C_3 fatty acids can be detected in the residue; this category includes residues composed of enough C_3 lipids that they were able to mask the presence of a small amount of C_4 compounds. A crucial corollary is that resources with more lipids tend to mask the presence of resources with fewer lipids.

This means that interpreting the overall amount of maize in a given residue is extremely difficult, because the interpretation is partially dependent on the other, unknown resources processed in the vessel. In fact, one of the peculiarities of residues from the midwestern and southeastern United States, compared with many reported from Britain and Europe, is the apparently greater abundance of primarily plant-based residues in the North American residues. This interpretation is based on the presence of plant biomarkers and on overall fatty acid composition, as described in chapter 2. After some research on the amount of lipids in resources commonly detected on archaeological sites, masking may be the cause of this quirk. Table 6.2 shows the average percentage of lipid by weight for a range of resources often processed in vessels both in Europe and Britain, and in the midwestern and southeastern United States. Although people in both North America and Eurasia regularly ate nuts, which are usually high in lipids, ethnohistoric accounts of many North American Indian groups observed that they used nuts to a degree that European observers thought was unusual (Swanton 1979). In addition, fat levels in many Eurasian domesticated animals are higher than in wild ones, largely because higher fat content is usually selected for during the domestication process. Wild meats used by North American peoples were generally lower in fats, with bear having the highest fat levels of the commonly hunted animals in the region. Conversely, many of the domesticated grains and seeds commonly used in Europe and Britain are relatively low in lipids, while those used in North America range from 0.84 to 51.9% lipid by weight, as can be seen in Table 6.2.

Combining these data, we can see that maize is actually fairly high in lipids for a domesticated grain and contains slightly more lipid/wt than the average of raw venison. It does contain noticeably less lipid/weight than a variety of other important resources in North America. If an equal amount of chenopodium, maize, and nuts were processed in a pot, the fatty acids from nuts would predominate and would tend to mask those from the maize in an isotopic analysis. If an equal amount of chenopodium, maize, and deer meat were processed in a pot, however, the maize would appear to be the dominant resource in the residue when tested isotopically, because it would not be masked by the other resources.

It therefore can be difficult to determine how much maize was originally processed in a pot by looking at isotopic analyses of the bulk residue or the fatty acids alone, even aside from issues of animals eating maize. In Figure 6.2, many of the fatty acids are more enriched than the known C_3 resources but are more depleted than the known C_4 ones, demonstrating mixture between fatty acids deriving from C_3 resources and those deriving from C_4 or CAM ones. In fact, in the original interpretation of the residues from the maize along the Mississippi project in 2000–2001, I originally interpreted many of the sherds whose fatty acids had $\delta^{13}C$ values between –25 and –23‰ as containing lipids from primarily C_3 resources. In 2001, very little basic CSIA had been done on wild resources. In the absence of most of the later baseline research, only fatty acids that clearly clustered near known C_4 values were interpreted as definitely containing a C_4 contribution. Now that more data have been collected on compound-specific carbon isotope values from a range of C_3 plants, C_4 plants, fish, and various terrestrial animals, it is clear that some of these values represent a blend of fatty acids from both C_3 and C_4 photosynthetic contributions. It is still not possible to determine what percentage of maize was present in the overall contents of a pot, because it depends on the lipid levels of resources originally processed in a pot, which can't usually be clearly determined without a biomarker.

TABLE 6.2. PERCENTAGE OF LIPID BY WEIGHT PRESENT IN SOME RESOURCES COMMONLY PROCESSED IN POTTERY FROM BRITAIN AND EUROPE AND FROM THE MIDWESTERN AND SOUTHEASTERN UNITED STATES

Resource	**% lipid/wt. (Average)**	**Resource Location**	**Reference**
Chenopodium (*Chenopodium berlandieri*)	0.84	North America	Assad-Bustillos et al. 2014; Gremillion 2004
Wheat (*Triticum spp.*)	1.7	Eurasia	Gremillion 2004
Maygrass (*Phalaris caroliniana*)	2.2	North America	Gremillion 2004
Mussel (*Bivalvia*)	2.2	Very widespread	USDA, Agricultural Research Service 2019
Chestnut[1] (*Castanea sp.*)	2.6	Most warm temperate zones, Americas and southern Europe	Sathe et al. 2008
Catfish, wild (*Ictalurus punctatus*)	2.8	North America	USDA, Agricultural Research Service 2019
Deer (raw)[2] (*Odocoileus virginianus*)	4.0		USDA, Agricultural Research Service 2019
Maize (*Zea mays*)	4.6	North and Mesoamerica	Gremillion 2004; Nuss and Tanumihardjo 2010
Turkey (*Meleagris gallipavo*) (domesticated)	5.6	North America	USDA, Agricultural Research Service 2019
Amaranth (*Amaranthus spp.*)	5.7	North America and Mesoamerica	Gremillion 2004
Salmon, wild (*Salmonidae*)	5.7	North America and Europe	USDA, Agricultural Research Service 2019
Bear (*Ursus sp.*)	8.3	North America	USDA, Agricultural Research Service 2019
Eel (*Anguilleformes*)	11.7	Western Europe, Asia	USDA, Agricultural Research Service 2019
Pork (raw, ground) (*Sus scrofa*)	21.2	Eurasia	USDA, Agricultural Research Service 2019
Beef (raw, average) (*Bos taurus*)	23.5	Eurasia	USDA, Agricultural Research Service 2019
Acorn (*Quercus spp.*)	23.9	Most temperate zones, Americas and Western Europe	Sathe et al. 2008

TABLE 6.2. CONTINUED

Resource	% lipid/wt. (Average)	Resource Location	Reference
Sumpweed (*Iva annua*)	44.5	North America	Gremillion 2004
Almond (*Prunis dulcis*)	47.0	Middle East	Sathe et al. 2008
Sunflower achenes (*Helianthus annuus*)	51.9	North America	Gremillion 2004; USDA, Agricultural Research Service 2019
Hickory nut (*Carya spp.*)	64.4	North America	Sathe et al. 2008
Walnut (*Juglans spp.*)	64.9	Most temperate zones, North America and Europe	Sathe et al. 2008
Pecan (*Carya illinoinensis*)	69.1	North America	Sathe et al. 2008

[1] Owing to the collapse of the American chestnut population (Castanea dentata), data are from the European chestnut tree (Castanea sativa); however, percentages should be similar.

[2] Lipid levels varied depending on cut and location. An average was taken of all raw venison.

An Example of Masking and Its Effects on Vessel Content Estimation

For example, we might extract a residue from a sherd with 15 μg lipid/g sherd. Isotopically, the fatty acids are –25.0‰ for both the $C_{16:0}$ and $C_{18:0}$, which are predominantly C_3 results, and which do not show clear evidence for C_4 components. The *n*-dotriacontanol, however, has a $\delta^{13}C$ value of –21.0‰, which shows a C_4 plant contribution, probably deriving from maize. In fact, given variation in the isotopic value of maize lipids, the OL_{32} could be composed almost entirely of maize contribution. The interpretation of the overall residue is of primarily plant resources, with meat resources also present. This interpretation is based on the presence of sitosterol, which is more abundant than cholesterol, and also on the fatty acid composition. This is predominantly unsaturated and with $C_{16:0}$ more abundant than $C_{18:0}$. As this is a hypothetical case, we'll make it easier by giving the cholesterol a C_3 value of –28.0‰ and the sitosterol another borderline value of –24.0‰. All the C_4 contribution in the residue therefore derives from plant resources. If we make certain assumptions, we can feed these numbers into a statistical isotopic mixing model and come up with an approximate result for the original contents of the pot (Fernandes et al. 2014). This looks amazingly scientific! But as with all statistics, it is dependent on the assumptions, which in this case are difficult to tease apart. Table 6.3 shows the different dietary results from the same statistical model (FRUITS 2.1.1) with different original content assumptions—all of which would be reasonable under most circumstances in a site in, say, Illinois.

TABLE 6.3. RESULTS OF MIXING MODELS FOR THE SAME RESIDUES BUT WITH A VARIETY OF REASONABLE RESOURCE ASSUMPTIONS

Assumed contents	$\delta^{13}C$ $C_{16:0}$, $C_{18:0}$	Modeled pot contents, standard deviation (SD)
maize, deer, generic nut	−25.0, −25.0	51% maize (SD 17) 31% deer (SD 21) 19% nuts (SD 13)
maize, deer, pecan nut	−25.0, −25.0	59% maize (SD 17) 30% deer (SD 21) 11% pecan (SD 6)
maize, deer, chenopodium	−25.0, −25.0	41% deer (SD 21) 39% chenopodium (SD 25) 21% maize (SD 10)
maize, catfish, sumpweed	−25.0, −25.0	52% maize (SD 17) 33% catfish (SD 22) 15% sumpweed (SD 8)
maize, catfish, chenopodium	−25.0, −25.0	43% catfish (SD 25) 40% chenopodium (SD 25) 17% maize (SD 9)
maize, mussel, chenopodium	−25.0, −25.0	44% mussel (SD 23) 39% chenopodium (SD 25) 17% maize (SD 10)
maize, average C_3 plant, average terrestrial and lacustrine resource	−25.0, −25.0	47% maize (SD 27) 22% C_3 plant (SD 15) 32% animal (SD 21)

Note: Program used was FRUITS 2.1.1, a Bayesian isotopic mixing model for dietary reconstruction (Fernandes et al. 2014).

As can be clearly seen from the data in Table 6.3, this is an unreliable way to model pot contents in mixed residues. Depending on the assumed original contents of the vessel, the amount of maize in the overall contents varies from 59% to 17%, and even within given assumption runs, the standard deviations for all resources are unacceptably high. Further, as this was a thought problem, several other sources of uncertainty were not included in the model. For example, each assumption included only three possible vessel contents, which is an arbitrarily small number. It is more realistic for a mixed residue to contain a large number of resources over its use-lifetime. Note that this simulation did not take into account any natural variation in $\delta^{13}C$ lipid values. Realistically, C_3 values of $C_{16:0}$ and $C_{18:0}$ vary from about −32 to about −25‰, and C_4 values of these fatty acids range from −22 to −16‰.

For the purposes of this calculation, we did not include this uncertainty, and arbitrarily set C_3 fatty acids at –28 and C_4 ones at –17, which are the averages in our database. In addition, in many midwestern contexts the $\delta^{13}C$ value of cholesterol in residues indicates that animals have been eating C_4 plants (presumably by browsing in the fields), and this changes all modeling calculations, making the presence of maize in the residue as a plant resource more uncertain. This is why the amount of maize in pottery residues cannot be calculated in a way that is really satisfying to excavators. There are too many variables present in the various isotopic and lipid contributions to residues and too little interpretive certainty about specific resources processed in a vessel. The best reasonable estimate is to say that there was a C_4 contribution to the residue, probably deriving from maize.

CSIA DETECTION OF MARINE ORGANISMS IN RESIDUES

Although considerable research has gone into the detection of maize in residues using CSIA, the analysis can be used in any situation where there are some resources that are isotopically different from others. For example, CSIA of fatty acids has been used to determine the amount of marine resources processed in pots in Alaska (Anderson et al. 2017; Craig et al. 2013, and others), the Baltic (Robson et al. 2019), islands in the northeastern portion of Europe, including Britain (Cramp et al. 2014), and a variety of other locations. Figure 6.2 showed $\delta^{13}C$ values of lipids excluding those from marine and anadromous resources; these values are now included in Figure 6.4. Because of a complex web of factors, CSIA lipid values for marine organisms overlap those for C_4 plants. The primary reason for this is that the source of carbon for marine organisms is not CO_2 from the air but rather dissolved carbon in the seawater, as briefly discussed. The various influences on the $\delta^{13}C$ values of this dissolved carbon are complex and include liquid water temperature, the amount of ice locked in the ice caps, water pH, and the amount of carbon held in sediments or living organisms at the time (Kohler et al. 2010). On average, however, the $\delta^{13}C$ value of marine carbon is about 7‰ more enriched than atmospheric carbon, which is generally around –7‰ (Chisholm et al. 1982), although there is variation by ocean depth and temperature. These values are passed up through the food chain by phytoplankton and marine algae, which will generally discriminate against ^{13}C similarly to plants to an extent somewhere between C_3 and C_4 photosynthesis on land. As a result, marine $\delta^{13}C$ values, both bulk and CSIA, are more enriched than those present in C_3 food webs on land and overlap those of C_4 food webs.

The range of values for marine organisms seen in Figure 6.4 probably reflects the entire complex series of factors involved in $\delta^{13}C$ values of oceanic carbon, including most notably water temperature and depth. It will also

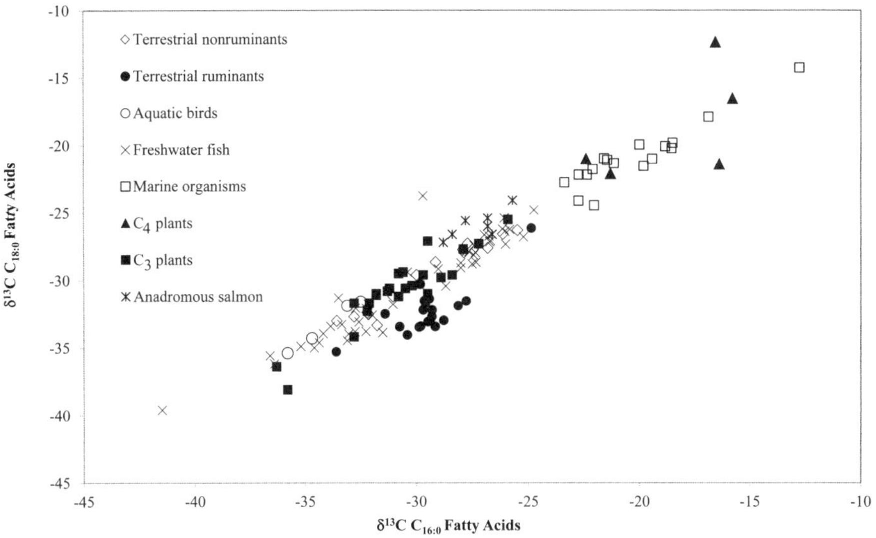

Figure 6.4. CSIA values for $C_{16:0}$ and $C_{18:0}$ for all the modern resources shown in Figure 6.2, but with the addition of marine and anadromous organisms (Ballentine et al. 1996; Choy et al. 2016; Craig et al. 2011, 2012, 2013; Fernandes et al. 2018; Spangenberg 2016; Spangenberg and Ogrinc 2001; Steele et al. 2010; Taché and Craig 2015; Woodbury et al. 1998). These resources overlap isotopically with C_4 plants. (Eleanora A. Reber)

include some shift caused by necessary lipid synthesis as needed by each separate marine organism, as previously discussed.

As a result, CSIA of lipids can be used to determine the presence of marine resources in pottery residues as long as it is unlikely for C_4 resources to also be present in the residues. For example, Karine Taché and Oliver Craig successfully identified marine fats in residues from Early Woodland Vinette pottery in the northeastern United States and southeastern Canada using CSIA of residues, among other techniques (Taché and Craig 2015). This was possible because there are very few C_4 or CAM plants potentially present in the diet in that area and time period. Similarly, Shelby Anderson and others could use enriched CSIA of fatty acids as an indicator for marine fats in the Arctic because there are no known C_4 or CAM plants in that area (Anderson et al. 2017). It is worth noting that this study concluded that freshwater fish rather than marine fish were primarily processed in the pottery; however, marine resources in that area would be the only potential source of fatty acids that are isotopically enriched in carbon. The absence of such enriched fatty acids with the presence of indicators for fish and shellfish (as discussed in chapter 7) allowed the interpretation of freshwater fish processing in the vessels.

CSIA identification of marine resources becomes more difficult in areas with both C_4 plants and marine resources. Sites in Polynesia would present interpretive difficulties when doing CSIA on fatty acids in residues, because both marine resources and sugarcane (a C_4 plant) are present. In this and other similar cases, the residue analyst will need to determine whether marine resources are present based on lipid interpretation, as described in chapter 7. CSIA of less common compounds, such as sterol biomarkers, alkanols, or highly unsaturated fatty acids or alkylphenyl fatty acids, would also be a good way to approach this problem. Enriched plant sterols would tend to point to a C_4 plant contribution to the residue, while enriched cholesterol or highly unsaturated fatty acids would tend to indicate a marine contribution.

CSIA DETECTION OF RUMINANTS AND MILK PRODUCTS

Many plant foods are low in fat compared with the animals that eat them. Thus, although all animals will "prefer" biochemically to remove fatty acids directly from the diet and use them without fractionation, some new fatty acids must be synthesized in the bodies of herbivores, usually from carbohydrates, as discussed. Ruminants, however, are able to digest more fatty acids from plants than are nonruminants, and this causes different CSIA values for their fatty acids than those in nonruminant herbivores. In a simplified form, ruminants are animals that digest tough plant matter by means of a sort of "pretreatment" of food in the rumen. This organ allows microbes to break down and ferment the tough plant tissue, releasing nutrients that are then digested. Much of the $C_{18:0}$ in ruminant fat derives from the microbial breakdown of more unsaturated 18-carbon fatty acids, such as $C_{18:3}$, $C_{18:2}$, and $C_{18:1}$; the remainder is synthesized from carbohydrates in the plant foods as usual. Because the ruminant can obtain more $C_{18:0}$ directly from plant foods than a nonruminant, there is less enrichment because of the synthesis of new lipids, and ruminant tissue fats are more depleted in $C_{18:0}$ than in $C_{16:0}$, allowing a slight but noticeable difference, shown in Figure 6.5 (Copley et al. 2003). The absorption of $C_{16:0}$ is also more efficient in ruminants, but not as much so as $C_{18:0}$ because highly unsaturated 18-carbon fatty acids are much more common than 16-carbon ones. This relative depletion can be used to determine whether an absorbed residue contains measurable amounts of ruminant fat, compared with nonruminant or plant fat. One way of measuring this difference is by subtracting the $\delta^{13}C$ values of $C_{18:0}$ from those of $C_{16:0}$ in the same residue. The result is a value sometimes called $\Delta^{13}C$, which in nonruminant lipids will normally be about zero. In ruminant body fat, this number seems to range from about −4.3 to −2.7 in wild ruminants (Craig et al. 2012) to −5.2 to −0.3 in domesticates (Copley et al. 2003; Evershed et al. 2002).

A similar calculation can be used to identify ruminant dairy products

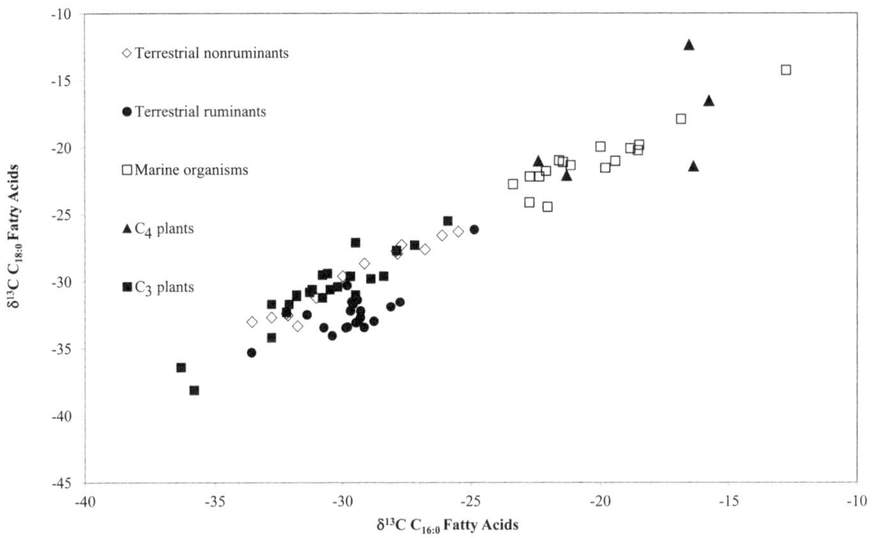

Figure 6.5. CSIA values for $C_{16:0}$ and $C_{18:0}$ for C_4 plants, C_3 plants, marine organisms, and terrestrial ruminants and nonruminants (Ballentine et al. 1996; Choy et al. 2016; Craig et al. 2011, 2012, 2013; Fernandes et al. 2018; Spangenberg 2016; Spangenberg and Ogrinc 2011; Steele 2010; Taché and Craig 2015; Woodbury et al. 1998). Note the relative depletion in ruminant fatty acids of $C_{18:0}$ compared with $C_{16:0}$. (Eleanora A. Reber)

in a residue. During milk production in the mammary glands, $C_{16:0}$ can be produced in the mammary glands from a mixture of dietary lipids and new synthesis, but $C_{18:0}$ cannot. Instead, it is incorporated both directly from the diet, as in body tissue, and is mobilized from $C_{18:0}$ already produced as the body fat, resulting in $C_{18:0}$ that derives more from microbial breakdown of unsaturated 18-carbon fatty acids. As a result, milk and dairy products from ruminants have $C_{18:0}$ that is even more depleted than the body fat of the ruminant that produced the milk when compared with $C_{16:0}$. For example, cow's milk is generally 2.3–2.9‰ more depleted than the body fat of the cow (Copley et al. 2003).

This technique has been most often applied in Europe, Great Britain, and the Near East where dairying made up an important part of the agricultural economy (Copley et al. 2003, 2005; Craig et al. 2007; Cramp et al. 2014; Dudd and Evershed 1998; Evershed et al. 2008a; Spangenburg et al. 2006, and others). It is important to note that each animal derives its milk lipids from its own diet and body fat. For example, the fatty acids of North American deer (a ruminant) are more depleted in $C_{18:0}$ than those from domesticated cattle eating a C_3 diet in Britain (Dudd and Evershed 1998; Taché and

Craig 2015). This is presumably because the deer are browsing on resources with more unsaturated 18-carbon fatty acids, permitting them to use more of these compounds directly from the breakdown of these lipids in their rumens. Alternatively, fatty acids from caribou, another ruminant, are regularly more enriched than those from all other ruminants. This is because of the importance of lichen in their diet, which as a symbiote is neither C_3 nor C_4 and has more enriched lipids than those in C_3 plants. In each region, however, ruminant dairy fats can be distinguished from ruminant body fats and nonruminant lipids. This is easier when these resources are not mixed in a single residue, but the presence of ruminant lipids of some sort can be detected isotopically, even in mixed residues.

THE BIG PICTURE

Isotope chemistry is an intricate topic, but harnessing it in absorbed residue analysis allows a large number of useful applications. As this technique becomes more widespread and the chemistry and biochemistry of food webs becomes better understood, more applications will probably be developed. Each of these applications will depend on the isotopic environment in each site area.

As described here, pottery that may contain both C_4 plants and marine organisms will be more difficult to interpret using CSIA of lipids than pottery that is likely to use only one of them. Although mixing curves can be constructed for pottery that contains both ruminant and nonruminant fat (Dudd and Evershed 1998), the more complex the isotopic variables going into each residue, the more difficult interpretation will become and the less helpful mixing models will be.

Despite these difficulties, however, the results of CSIA analysis of residues are very helpful in interpreting human behavior, permitting the detection of important resources in residues that otherwise cannot be detected using biomarkers or fatty acids alone. When in doubt about the interpretability of CSIA of lipid residues, it is often helpful to think about the interpretability of $\delta^{13}C$ values of skeletal analyses from the same site or region. The skeletal results will include everything eaten by an individual, including things cooked in pottery, so it is likely that an area with interpretable skeletal results will also have interpretable results for CSIA of pottery residues. If, conversely, $\delta^{13}C$ values of skeletal values are not helpful in the area where you are digging, then only relatively subtle isotopic interpretations will be possible when analyzed in residues, such as those deriving from ruminant fats and dairying.

7

FISH AND SHELLFISH DETECTION

The detection of fish and shellfish has been of considerable interest to residue analysts and archaeologists over the years. Like starchy grains and nuts, fish and shellfish can be difficult to detect directly in the archaeological record, although archaeologists know they were an important part of the diet. Recently, advances in residue analysis have improved the detection of fish and shellfish in residues.

Compound-specific isotope analysis (CSIA) of residues can detect marine fish and shellfish in residues reliably, as discussed in the previous chapter. Freshwater resources, however, pose a challenge for residue analysts. Fatty acids in fish and shellfish are normally highly unsaturated and therefore do not preserve well in archaeological residues. In addition, many fish and shellfish are relatively low in fats and are therefore heavily masked. Although there are several useful indicators for fish and shellfish lipids in residues from both marine and freshwater environments, there is to date no distinctive biomarker.

In lieu of one specific biomarker or approach, several different indicators can be used to detect the presence of fish and shellfish resources. Each of the indicators, however, has some potential difficulty that renders it less than perfect. Table 7.1 summarizes these indicators and their uses and potential limitations. In general, though, fish and shellfish detection in residues is more accurate when more indicators are present in a given residue. Definitive identification of freshwater fish or shellfish lipids can only be made in the presence of at least two indicators; more than two is optimal.

This chapter discusses these indicators in detail, including lipid contents and the chemistry of most fish and shellfish. A case study on some residues from Angel Mounds explains how several undramatic indicators of fish and/or shellfish lipids permit definitive identifications of freshwater resources—as

Table 7.1. Indicators for Fish and Shellfish Lipids in a Residue with Potential Strengths and Weaknesses in Approximate Order of Reliability

Indicator	Strengths	Weaknesses
Compound-specific isotope analysis (discussed in chapter 6)	Very reliable when detecting marine resources in an environment with no C_4 plants	Unreliable on freshwater resources; can get confusing in environments with both marine and C_4 resources
Polyunsaturated fatty acids (PUFAs)	Present in both marine and freshwater resources; more common and abundant in these resources than almost any other. A reliable indicator, when found	Very poor preservation in the archaeological record. Unlikely to survive in a residue
ω-(o-alkylphenyl) alkanoic acids (APHs)	Formed by heating polyunsaturated fatty acids (PUFAs) to 270–300°C. APHs with 20 or more carbons are a strong indicator for fish or shellfish lipids	Only form at high temperatures. Fish lipids processed at lower temperatures will not have this biomarker. APHs with 16 or 18 carbons could derive from high-temperature heating of plant PUFAs
Isoprenoid fatty acids	Known to occur in many marine fats, likely to occur in freshwater fish and shellfish fats. Preservation good, extraction better with direct acidic extraction methods	Poor recovery with chloroform/ methanol extractions. They also occur in resources other than fish and shellfish. Disagreement over how distinctive these compounds are to fish and shellfish
More abundant long-chain monounsaturated fatty acids ($C_{20:1}$, $C_{22:1}$, $C_{24:1}$)	Useful indicator for fish/ shellfish, as these compounds are often more abundant in fish/ shellfish lipids than other lipids	Compounds are also present in many plant lipids. Usually they are present in small amounts but in some cases can be very abundant
$C_{18:0}/C_{16:0} < 0.48$, + cholesterol	Works best on residues derived almost entirely from fish/shellfish lipids	Easily masked by lipid-rich resources in a residue. Might not work on residues that underwent extreme microbial degradation

long as the residue analyst notices the indicators. Many of the distinctive components in fish and/or shellfish lipids are highly branched or polyunsaturated. At least some of these components appear to bind tightly to the ceramic matrix in absorbed residues and become part of the bound insoluble fraction for chloroform/methanol solvent extraction, as discussed in chapter 2. As a result, detection of fish and shellfish lipids has increased as acidic and alkali extraction techniques gain in popularity. This chapter includes another case study from George Reeves to examine the effect of extraction methods on the interpretation of fish and/or shellfish in residues.

COMPOUND-SPECIFIC ISOTOPE ANALYSIS IN MARINE AND FRESHWATER RESOURCES

CSIA analysis can be invaluable in detecting marine resource contributions to a residue if these resources make up an appreciable part of the diet, and if there are no C_4 or CAM photosynthesizing plants in the area. This is because of the isotopically different carbon food web present in oceans compared with that in terrestrial food webs. If the marine resources make up only a small part of the diet, the relatively low-fat marine resources are likely to be masked in residues by higher-fat terrestrial resources. If C_4 or CAM photosynthesizers are present in the area, the enrichment in $\delta^{13}C$ values of lipids could derive from plant contributions to the residue, instead of the marine resources. As described, CSIA of sterol biomarkers and neutral compounds can be used to disentangle the possible sources of the enriched carbon. In some cases, however, isotopes will not be as helpful as we might want, and lipid interpretation must be used to detect fish and shellfish processing.

CSIA cannot usually be used to detect lipids from freshwater fish and shellfish. This is because the primary source of most freshwater carbon is terrestrial, deriving from runoff or detritus falling into the stream or lake. Carbon isotope values from freshwater fish and shellfish therefore usually are identical to values from terrestrial resources.

An exception to this rule exists if the freshwater organisms live in lakes or rivers that are in regular contact with limestone bedrock, or that have undeveloped soils (Philippsen 2013). In this case, the carbon sustaining the food web will derive from dissolved limestone, which derives from marine creatures. As a result, carbon from freshwater organisms in these environments will derive from predominantly marine resources, and their $\delta^{13}C$ values will appear similar to those of marine fish and shellfish.

This effect is related to the freshwater reservoir effect (FRE) for radiocarbon dates. The FRE produces older dates in some freshwater resources than in contemporaneous terrestrial resources if there is old dissolved carbon in the water from a variety of origins, including limestone bedrock, melting glaciers, and such (Fischer and Heinemeier 2003; Hart et al. 2013; Keaveney and Reimer 2012; Philippsen 2013). Interestingly, marine $\delta^{13}C$ values for freshwater resources do not seem to correlate directly to radiocarbon FRE effects, with stable isotope values of bone collagen apparently affected more by the presence of well-developed soil at the base of lakes and rivers than dissolved carbonates in the water (Philippsen 2013). So the mere presence of a

FRE in radiocarbon dates from fish or shellfish in a lake doesn't necessarily mean that the stable isotopes of these fish or shellfish will be distinguishable from terrestrial. In addition, there is a good deal of variation in both FRE and stable isotopes caused by a variety of factors, including the presence of recent rainfall and the individual feeding habits of the species of organism being tested (Keaveney and Reimer 2012; Philippsen 2013). The upshot of all this complexity is that it is unwise to depend on CSIA to detect freshwater resources in a residue. In these circumstances, the residue analyst must rely on lipid interpretation without the aid of isotopes.

FATTY ACID COMPOSITION IN FISH AND SHELLFISH

The fundamental difficulty with identifying fish and shellfish lipids is that they are distinctive but not compositionally unique. Most indicators for fish and shellfish lipids are essentially identifying these distinctive traits of fish or shellfish lipids. These factors are not unique, so they do not serve as biomarkers.

As is common for all creatures, fish and shellfish prefer to use fatty acids directly from their diet rather than synthesizing them from other foods. An important difference between most fish/shellfish lipids and those from terrestrial animals, however, is that most fish and shellfish ultimately derive their fatty acids from algal lipids at the base of the freshwater food chain (Harwood 2019) or from phytoplankton at the base of the marine food web. Both sources contain notably higher incidences of longer, more unsaturated fatty acids than are normally present in plant lipids. Although the most abundant fatty acids in fish and shellfish are saturated or monounsaturated, there is a much higher incidence of longer and more unsaturated fatty acids in fish and shellfish than in terrestrial animals. Because of this, more abundant unusual monounsaturated fatty acids, such as $C_{20:1}$, and highly polyunsaturated fatty acids (PUFAs) are often viewed as distinctive features of fish and shellfish lipids. It is important to note that these distinctive lipids are also occasionally present in plant lipids.

Polyunsaturated Fatty Acids (PUFAs)

Among the most common PUFAs in fish and shellfish are $C_{22:6}$, $C_{22:5}$, $C_{20:5}$, $C_{20:4}$, $C_{18:4}$, and $C_{18:3}$ (Ackman 1964; Bragagnolo and Rodriguez-Amaya 2001; Evershed et al. 2008b; Gunstone et al. 1986; Lazzara et al. 2012, and others). Although many of these compounds can be found in small amounts in some plant resources, they are rare and occur in noticeably smaller amounts than in fish and shellfish (Gunstone et al. 1986). PUFAs, however, do not preserve well in the archaeological record because the large number of unsaturated bonds are especially prone to oxidation during burial. It is rare for such compounds to survive in the archaeological record, even in cases where preservation would be expected to be extraordinarily good (Brown and Heron 2005; Corr et al. 2008). As a result, for many years the detection of fish and shellfish resources in residues was difficult because the compounds that were especially unique to these resources did not survive in the archaeological record.

Alkylphenyl Fatty Acids (APHs)

In 2004, a team of researchers based at Richard Evershed's laboratory at the University of Bristol published several papers that tentatively identified process biomarkers that formed during the heating of PUFAs (Copley et al. 2004; Hansel et al. 2004). Upon investigating residues from pottery originating from both Brazil and South Africa that might be expected to have processed fish, the investigators noticed a series of unusual alkylphenyl fatty acids in both. Alkylphenyl fatty acids (APHs; ω-(o-alkylphenyl)alkanoic acids in the scientific form) are fatty acids that contain a stable aromatic ring in the middle of the lipid chain as shown in Figure 7.1. The presence of an aromatic ring makes these compounds unusually stable and persistent in residues. These compounds do not occur naturally but were originally observed forming from the heating of PUFAs (Matikainen et al. 2003; Michael 1966; Michael et al. 1966). The residue analysts suggested that a similar reaction might occur in pottery residues if PUFAs are present in a residue and the

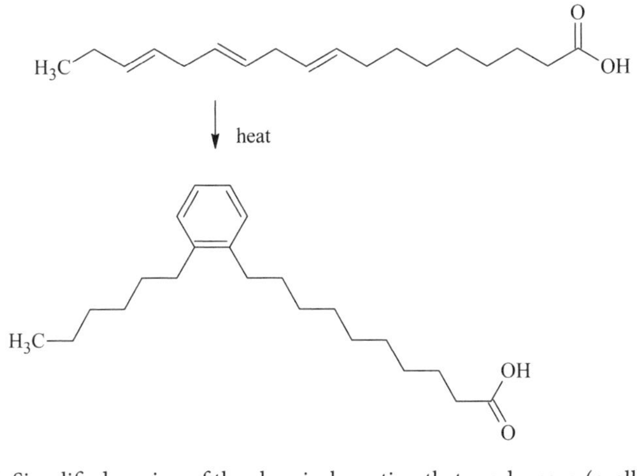

Figure 7.1. Simplified version of the chemical reaction that produces ω-(o-alkylphenyl) alkanoic acids (APHs) from polyunsaturated fatty acids with at least three double bonds, in this case, linoleic acid, $C_{18:3}$. The aromatic ring in the APH makes the compound much more stable and persistent in residues than the precursor fatty acid. More details are given in other publications (Copley et al. 2004; Evershed et al. 2008b; Matikainen et al. 2003). (Eleanora A. Reber)

residue heated at temperatures above 270°C (Evershed et al. 2008b). The result is the production of a stable biomarker produced from the unstable PUFAs in fish, but only if the residue was heated at some point to a high temperature. Most simmering or cooking does not reach temperatures much above 100°C, if that. Temperatures at or above 270°C would require either a cooking method like frying or dry toasting or that the sherd was accidentally heated to high temperatures during use or after breakage.

This suggestion was confirmed experimentally in 2008 (Evershed et al. 2008b), and APHs have since been detected in various residues around the world at a relatively low incidence. The process biomarkers appear to be useful indicators for lipids derived from fish or shellfish, but it is rare for residues to be heated to the required temperatures to produce the biomarkers.

Note: Do APHs Always Indicate Fish or Shellfish?

Not always. The production of these compounds requires an unsaturated fatty acid with at least three double bonds, as shown in Figure 7.1. Such fatty acids, especially with 16- and 18-carbon chains, occur in plant lipids as well as fish and shellfish. They don't normally appear in residues for the same reason that PUFAs don't appear in fish/shellfish residues: they break down too quickly to survive in the archaeological record. For this reason, the papers describing these process biomarkers added several additional conditions to confirm the presence of fish and/or shellfish lipids (Copley et al. 2004; Evershed et al. 2008b; Hansel et al. 2004). First, fish and shellfish lipids are generally longer than plant lipids. So APHs deriving from 16 or 18 carbons are more likely to derive from plants than those with 20, 22, and 24 carbons. If a residue contains 16- and/or 18-carbon APHs, but not those with longer carbon chains, the compounds may have derived from high temperature heating of plant lipids. Residues with 20–24 carbon APHs are much more likely to derive from fish or shellfish lipids. Ideally, these compounds would be accompanied by one of the other indicators of fish and shellfish discussed in this chapter. Even in the absence of a secondary confirmation, however, the presence of 20–24 carbon APHs is considered a probable marker for fish and/or shellfish processing in the pottery. In addition, the presence of any APHs indicates that the residue was heated to at least 270°C, which can be used to interpret either how the vessel was used to cook or whether it was accidentally burned at some point.

Isoprenoid Fatty Acids

Another helpful indicator for the presence of fish and shellfish lipids in a residue is the presence of isoprenoid fatty acids. These fatty acids are not

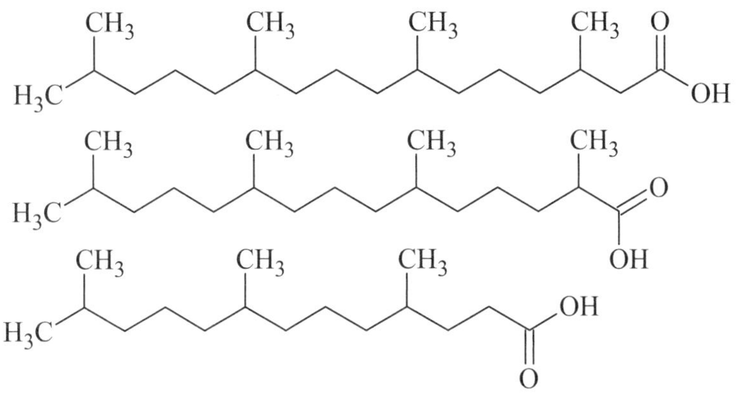

Figure 7.2. Three isoprenoid fatty acids commonly used as indicators for fish or shellfish presence in a residue. Phytanic acid is on the top, pristanic acid in the center, and 4,8,12 trimethyltridecanoic acid (TMTD) on the bottom. (Eleanora A. Reber)

polyunsaturated but do show an unusual incidence of branches in a regular pattern, as shown in Figure 7.2. Three isoprenoid fatty acids are often used as biomarkers for fish and shellfish resources—phytanic acid, pristanic acid, and 4,8,12 trimethyltridecanoic acid (TMTD). As these compounds are saturated rather than polyunsaturated, they preserve better in residues. They are also found in unusually abundant amounts in marine lipids (Ackman and Hooper 1970; Ackman et al. 1971; Evershed et al. 2008b, and others). For years, these compounds were relatively rare, only detected in situations with unusually good lipid preservation (Corr et al. 2008). Fortunately, as direct acidic and alkali extraction methods have increased in popularity, isoprenoid fatty acids have been much more regularly detected and are often used as an indicator for the presence of fish and shellfish resources (Craig et al. 2013; Farrell et al. 2014; Lucquin et al. 2016; Taché and Craig 2015, and others). The reason for this increase seems to be that the isoprenoid fatty acids are more tightly linked to the clay particles in pottery than straight-chain fatty acids. Chloroform/methanol extraction, therefore, does not seem to extract isoprenoids as efficiently as acidic and alkaline extractions (Correa-Ascencio and Evershed 2014; Craig et al. 2004).

Further research is needed on the occurrence of isoprenoid fatty acids in various resources. It is clear that these isoprenoid fatty acids are unusually abundant in both marine and freshwater fish and shellfish (Ackman and Hooper 1970; Ackman et al. 1971; Avigan and Blumer 1968; Lucquin et al. 2016). These compounds can therefore be used as a strong indicator for the presence of fish or shellfish lipids. That said, these compounds appear to derive from the metabolic or microbial degradation of chlorophyll (Gunstone et al. 1986; Verhoeven et al. 1998) and therefore can be present in a wide range of resources. These include ruminant fat, ruminant milk, and human lipids (fingerprint oil). In geology, the compounds are sometimes used as general biomarkers for plant breakdown (Brassell et al. 1983; Maxwell et al. 1973). These compounds are a useful indicator for fish or shellfish lipids, but they are not biomarkers for these resources. This is especially the case in mixed residues that contain potential evidence for ruminant fats or dairy products.

Monounsaturated Long-Chain Fatty Acids

As mentioned, long-chain monounsaturated fatty acids $C_{20:1}$ and $C_{22:1}$ appear to be more common in many fish lipids than in most plant lipids (Brown and Heron 2005; Gunstone et al. 1986). The compounds aren't true biomarkers, however, as they occur in many plants, usually in small amounts. A residue with measurable surviving amounts of $C_{20:1}$ and $C_{22:1}$, especially if these amounts are equivalent to or larger than the amounts of $C_{18:1}$ in the residue, suggests the presence of fish or shellfish lipids but does not establish it beyond doubt.

$C_{18:0}$ / $C_{16:0}$ < 0.48 + CHOLESTEROL

A further indicator was suggested by Monika Olsson and Sven Isaksson (2008). They argued that fish oil would tend to be dominated by $C_{16:0}$, similar to a plant residue, but would contain large amounts of cholesterol. Therefore, fish oil could be indicated by a ratio of $C_{18:0}$ / $C_{16:0}$ of 0.48 or less, with the presence of cholesterol. Without the presence of cholesterol, such a ratio would tend to indicate plants, or be a more generic "primarily plant or fish/shellfish" indicator. This indicator is most useful when a residue is derived almost completely from fish or shellfish lipids. In mixed residues, mixing of fatty acids would tend to move the fatty acid ratio away from that typical of fish and shellfish. Masking is also a problem with this indicator, as a residue containing large amounts of fatty acids from fish or shellfish that are low in fat could be masked by the presence of relatively small amounts of terrestrial resources that were high in fat. As with all the indicators for fish and shellfish in a residue, except CSIA analysis, the analyst will need two or more positive indicators for a confident identification of fish or shellfish lipids in a residue.

Case Study: Freshwater Fish and Shellfish in Pottery from Angel Mounds

The process of detecting fish or shellfish residues in residues is easier than it sounds, as long as the analyst remembers the possibilities. A good example of this is a series of residues extracted from pots at Angel Mounds, an

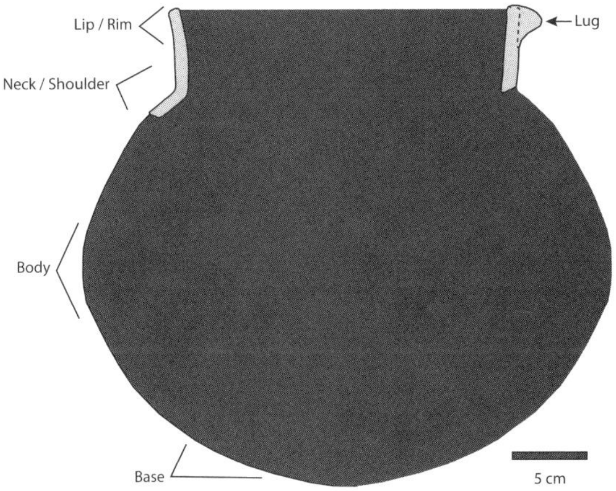

Figure 7.3. Silhouette of the Mississippi Plain jar from Angel Mounds and the locations where it was sampled. (Eleanora Reber, Timothy E. Baumann, G. William Monaghan, and Kelsey Noakes Meyers, Absorbed Residue Analysis of Mississippi Plain Jar from Angel Mounds [12Vg1]: Lipid Distribution Revisited. *Advances in Archaeological Practice* 3[1]:29–49, 2015, reproduced with permission)

important Mississippian site located along the banks of the Ohio River in Indiana. Among the residues analyzed from the site were four separate residues extracted from different locations on a single complete Mississippi Plain jar excavated from a burnt house at Angel Mounds, and shown in Figure 7.3 (Reber et al. 2015). Four samples were taken from different parts of the same vessel as a test on cooking practice and lipid absorption (cf. Charters et al. 1993a). A helpful side effect of this sampling procedure, however, was that it allowed duplication and confirmation of results from a single pot, even when the interpretations were complicated. The other three residues from pottery in the project were extracted from three different negative-painted pottery (NPP) vessels. Eventually, three of the four vessels sampled showed indicators for fish and shellfish contribution: two of the three negative-painted sherds, and the Mississippi Plain jar. Samples of soil interior and exterior to the complete jar were taken. An earlier study suggested that when a complete pot is buried with contents intact, many of the residues from the foodstuffs remain soaked into the soil that falls into the complete vessel during burial (Reber et al. 2010).

When the sherds were originally extracted using the solvent extraction methodology and analyzed, in 2010, I immediately noticed the unusually good preservation of triacylglycerols, waxes, and other complete lipids

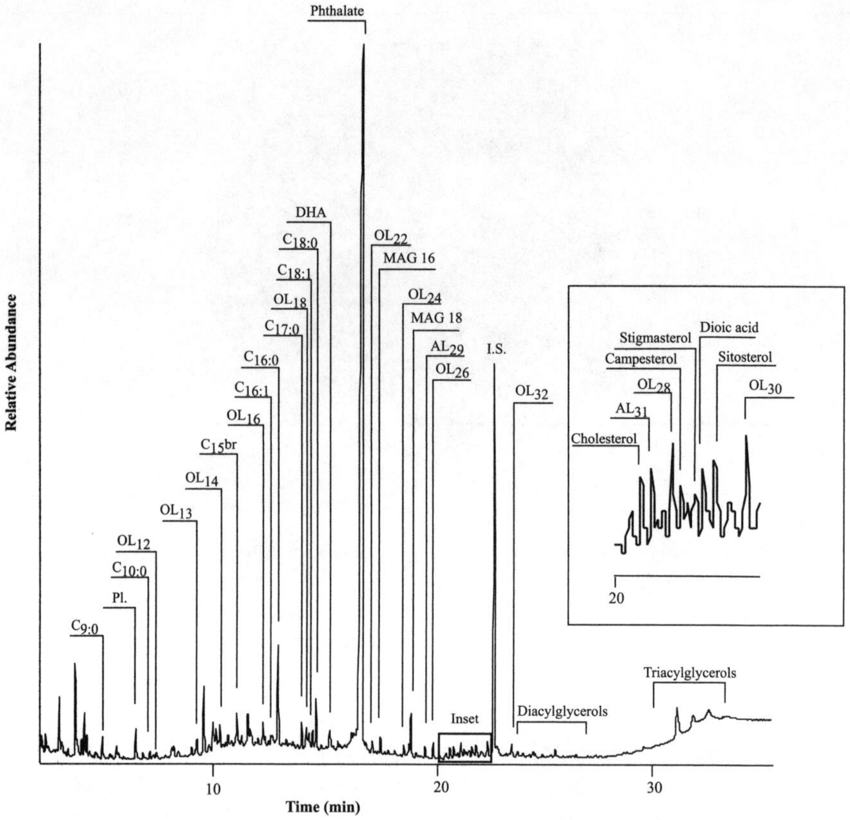

Figure 7.4. Total lipid extract of the residue extracted from the shoulder of the Mississippi Plain jar with chloroform/methanol extraction. Note the good preservation of triacylglycerols. (Eleanora A. Reber)

(Figure 7.4). In contrast, there were relatively few free fatty acids for an archaeological residue. The lipids looked almost intact, with relatively minimal microbial breakdown. Cholesterol was present in every residue in the study, and there was a wide range of alkanols and alkanes. The residues from the NPP sherds contained no plant sterols, while all residues from the complete plain jar contained a range of plant sterols. The NPP residues contained a range of diterpenoids and waxes, which probably derived from the negative-painting process. This decoration was probably produced by rubbing plant resins and waxes over the parts of the vessel that were meant to fire with a black color (Baumann et al. 2013). Most of the residues in the study were originally interpreted to be a mixture of plant and meat resources, including the residues from the Mississippi Plain jar. This interpretation was based

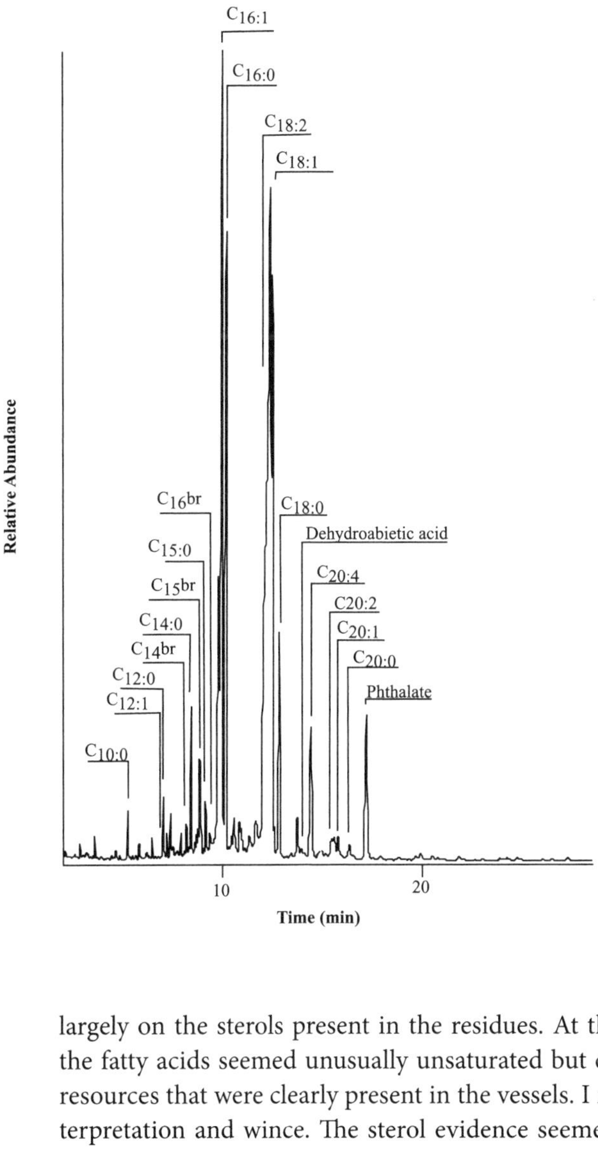

Figure 7.5. Fatty acid fraction of the residue extracted from the base of the Mississippi Plain jar. Note the highly unsaturated fatty acids and presence of 20-carbon PUFAs. There is also some plasticizer contamination, shown by the phthalate. (Eleanora A. Reber)

largely on the sterols present in the residues. At the time, I did notice that the fatty acids seemed unusually unsaturated but chalked it up to the plant resources that were clearly present in the vessels. I now look back on this interpretation and wince. The sterol evidence seemed clear, the alkanols and alkanes were all indicative of plants, but I was missing some important interpretive data.

In 2012, we began working on publishing the data. With two years' additional experience, I quickly realized that I'd missed a lot. The original interpretation noted that the fatty acids were unusually unsaturated (shown in Figure 7.5 and Table 7.2) but had assumed that it was due to unusually good preservation of plant residues. The most abundant fatty acid in most of the residues from the Mississippi Plain jar was $C_{18:1}$. In several cases, $C_{16:1}$ was

Table 7.2. Percentage of Compounds Found in Fatty Acid Fractions from Residues Extracted from Pottery at the Angel Mounds Site

	Negative-painted			Mississippi Plain jar			
Compound	RL 140	RL 142	RL 143	RL 144	RL 145	RL 147	RL 148
$C_{8:0}$	-	-	2	-	-	-	-
$C_{9:0}$	-	1	5	-	-	-	-
$C_{10:0}$	2	-	2	-	-	-	-
$C_{11:0}$	-	-	1	-	-	-	-
$C_{12:0}$	2	2	2	-	1	1	-
$C_{13:0}$	-	1	1	-	-	-	-
$C_{14:1}$	-	4	3	-	-	-	-
$C_{14:0}$	4	6	6	2	1	1	2
$C_{15:1}$	-	1	1	-	-	-	-
$C_{15:0}$	2	5	3	1	1	1	1
$C_{16:1}$	5	17	15	30	18	21	25
$C_{16:0}$	18	19	15	20	23	12	15
$C_{17:1}$	-	2	1	-	-	-	2
$C_{17:0}$	-	1	1	3	-	1	1
$C_{18:3}$	-	-	-	2	-	-	-
$C_{18:2}$	-	1	2	12	13	11	14
$C_{18:1}$	32	10	10	-	32	35	28
$C_{18:0}$	9	8	7	9	6	4	4
$C_{19:1}$	-	-	-	1	-	-	-
$C_{19:0}$	-	-	-	1	-	-	-
$C_{20:4}$	-	-	-	4	1	3	2
$C_{20:2}$	-	-	-	-	-	1	-
$C_{20:1}$	-	1	-	-	-	1	-
$C_{20:0}$	-	1	-	1	-	-	-

actually more abundant than $C_{16:0}$, which is very unusual and could actually be a warning sign for some form of modern contamination. Furthermore, every one of the residues from the Plain jar contained the PUFA $C_{20:4}$ in amounts ranging from 1 to 4% of the total fraction. The residue extracted from the shoulder of the Plain jar, generally believed to be the most abundant and interpretable location for sampling in cooking pots (Charters et al. 1993a), also contained at least 1% $C_{24:1}$, $C_{20:2}$, and $C_{20:1}$. Although there was

TABLE 7.2. CONTINUED

	Negative-painted			Mississippi Plain jar			
Compound	RL 140	RL 142	RL 143	RL 144	RL 145	RL 147	RL 148
$C_{22:0}$	-	-	-	1	-	-	-
$C_{24:1}$	-	-	-	-	-	1	-
$C_{24:0}$	-	1	1	1	-	-	-
C_{22} hydroxy	-	-	-	1	-	-	-
C_{13}br	-	1	1	-	-	-	-
C_{14}br	-	-	-	-	-	1	-
C_{15}br	1	8	6	2	1	2	3
$C_{16:1}$br	2	-	-	-	-	-	-
C_{16}br	1	1	1	-	-	-	-
C_{17}br	2	3	3	3	1	1	2
C_{18}br	-	-	-	1	-	-	-
C_{20}br	1	-	-	-	-	-	-
Phenylhexa-noic acid	-	3	5	-	-	-	-
5a-podocarpa C_{17}br6,11, 13trien16enoic acid	1	-	-	-	-	-	-
Isopimaric acid	4	-	-	-	-	-	-
Hydroxy-dehydroabietic acid	-	-	1	-	-	-	-
7-oxodehydro-abietic acid	-	-	1	-	-	-	-
Dehydro-abietic acid	3	-	-	1	-	-	-

Note: Compounds are listed by class and then by chain length.
RL = residue lab.

some plasticizer in the residues, none of them contained squalene or markers for sunscreen, insect repellent, or modern lotion. This suggested that the unusual fatty acid profile did not derive from modern contamination.

Eventually, I figured it out. The residues from the Mississippi Plain jar contained three supporting indicators for fish/shellfish lipids; I hadn't realized it at first because none of them, by themselves, were particularly dramatic. The combination of indicators together made the identification unusually definitive. Going back through our list of indicators in Table 7.1, it all seems obvious. First, and the thing I should have caught in the first interpretive round, was the presence of PUFAs. $C_{20:4}$ is present in some plant lipids, but it's very unusual. It is, however, a well-known and abundant compound in fish and shellfish residues and the third-most-abundant fatty acid in freshwater mussels (Newton et al. 2013:Table 3).

Once the presence of PUFAs started me looking at the indicators of fish and shellfish, it was clear that several others were present as well. Isoprenoid fatty acids were not present, nor were APHs. But more-abundant long-chain monounsaturated fatty acids were clearly present, as can be seen in Table 7.2. When the $C_{18:0}$ / $C_{16:0}$ ratio of fatty acids in the residues was calculated, the value was below 0.48 in all of the Mississippi Plain residues. As you'll recall, cholesterol was present in every pottery residue in the study, and can be seen in Figure 7.4. The Mississippi Plain vessel therefore had three supporting indicators for fish/shellfish. None of them was definitive on its own, but the three together were convincing evidence that the residue derived largely from fish and shellfish, with some plant contribution (plant sterols and alkanols).

The residues in the NPP sherds from the site were more complicated to interpret. This is partially because some of the compounds in their residues derived from the negative-painting process, as described. The contents of the NPP sherds also varied from the Mississippi Plain jar, however. The Mississippi Plain jar clearly contained plant resources as well as fish or shellfish, while the NPP residues contained no plant sterols and fewer alkanols and alkanes indicative of a plant contribution. In addition, the NPP residues did not contain any $C_{20:4}$, although two of them did contain $C_{18:2}$. However, $C_{18:2}$ and other 18-carbon PUFAs are known to be abundant in plant resources, while 20-carbon polyunsaturates are much more abundant in fish and shellfish. The most convincing evidence for freshwater fish or shellfish contribution to residues was therefore present in the Mississippi Plain jar residues, but absent from those from the NPP sherds.

The NPP sherds are therefore a more typical example of the sorts of interpretation that tend to happen when identifying a potential fish or shellfish contribution to a residue. These residues did not have convincing PUFAs, or isoprenoid fatty acids, or APHs. One of them, from a bowl sherd, did have $C_{20:1}$ making up 1% of the total fatty acid fraction. Two of the NPP sherds, the bowl and the bottle, had $C_{18:0}$ / $C_{16:0} < 0.48$, and also contained cholesterol. The third NPP, from a plate, had a $C_{18:0}$ / $C_{16:0} = 0.5$, which was just barely too high to fit the proposed indicator. One of the NPP sherds therefore had two indicators for fish or shellfish contribution, one had one indicator, and one just missed all criteria. The interpretation of the uses of these vessels is therefore more complicated than that of the Mississippi Plain jar.

At the end of the day, the NPP bowl could be interpreted as probably containing freshwater fish and/or shellfish lipids because its residues had two of the possible indicators. The NPP bottle, with one indicator of fish and/or shellfish lipids, could be interpreted as possibly containing fish or shellfish lipids. The NPP plate was interpreted as probably containing a mixture of animal and pine resin contributions. In truth, this sherd may have contained fish or shellfish, mixed in with other resources, based on the still unsaturated fatty acid abundances that were dominated by $C_{16:0}$. But in the absence of any indicators for fish or shellfish contribution, it just wasn't a tenable interpretation.

Case Study: Isoprenoids and Extraction Methodology—George Reeves

A different case study for fish/shellfish interpretation is furnished by the George Reeves site. In this case, we ran parallel extraction procedures on the same sherds to do a baseline comparison of the chloroform/methanol and acidic methanol extraction techniques. The result was a clear illustration of how residue extraction methodology can affect the interpretation of vessel contents, especially fish or shellfish lipids.

A comparison of the results of two sherds is shown in Table 7.3. Because of complexities in the sample naming process that we used for this study, we will simply label the two sherds 1 and 2. Using the TAH fractions, both sherds 1 and 2 are interpreted as containing fish and/or shellfish lipids. Sherd 2 has the strongest interpretation for fish/shellfish. It does not have PUFAs, but does have two separate isoprenoids, pristanic and phytanic acids. It also has the monounsaturated fatty acids $C_{20:1}$ and $C_{24:1}$, although in relatively small amounts—less than 1% of the total fatty acid fraction. The $C_{18:0}$ / $C_{16:0}$ is quite low, at 0.7, but not low enough to qualify for the ratio and cholesterol method, although cholesterol is present in the residue. The residue has two of the indicators for fish and/or shellfish lipids and can therefore be interpreted as likely containing fish and/or shellfish. It is not as strongly confirmed as the Mississippi Plain jar already discussed, which had three of the indicators.

The TAH residue from sherd 1, shown in Figure 3.4, also has two of the indicators for fish or shellfish, though not as strongly as sherd 2. The TAH fraction contains no PUFAs, and one isoprenoid fatty acid, phytanic acid, as opposed to the two isoprenoids found in the sherd 2 residue. It contains larger amounts of the monounsaturated fatty acid $C_{22:1}$ than that present in sherd 2, and might have contained $C_{20:1}$, except that plasticizer contamination made this portion of the spectrum unreadable. Unlike the TAH of sherd 2, this fraction does not contain any $C_{24:1}$. The $C_{18:0}$ / $C_{16:0}$ ratio of sherd 1,

Table 7.3. Relative Fatty Acid Abundances from Two Sherds Extracted Using Two Extraction Techniques

Compounds	Sherd 1 TAH	Sherd 1 FAME	Sherd 2 TAH	Sherd 2 FAME
$C_{14:0}$	2	1	-	-
$C_{16:1}$	2	5	-	-
$C_{16:0}$	26	32	43	
Pristanic acid	-	-	1	-
$C_{18:1}$	7	22	4	
$C_{18:0}$	56	35	32	
Phytanic acid	1	-	1	-
$C_{20:1}$	-	-	< 0.5	-
$C_{20:0}$	1	-	6	-
$C_{22:1}$	2	-	-	-
$C_{22:0}$	-	-	4	-
$C_{24:1}$	-	-	< 0.5	-
$C_{24:0}$	< 0.5	-	5	-

Note: Fatty acid abundances in the total acidic hydrolysis (TAH) fractions are made more comparable with those in the fatty acid methyl ester (FAME) fractions by calculating only from fatty acids in the TAH fraction.

however, is noticeably closer to that for a residue derived primarily from meat, at 2. Both the TAH fractions from sherds 1 and 2 are interpreted as containing a mixture of foodstuffs, probably including fish and/or shellfish.

An interesting sidelight shines on the residues when we compare the interpretation of the chloroform/methanol extractions with the acidic methanol extractions from the sherds. Acidic methanol extraction permitted the interpretation of possible fish and/or shellfish resources in the residues, which could not be obtained using the chloroform/methanol extraction.

Sherd 2 could not be interpreted at all using the chloroform/methanol extraction, because it extracted insufficient residue for interpretation. This is presumably because the acidic methanol extraction removed lipids from the ceramic matrix that were bound through physicochemical means tightly to the ceramic and that were not removed by the chloroform/methanol. Most of the residue in the vessel walls that could easily be extracted from the ceramic had already been removed, probably either through hydrolysis or microbial breakdown, leaving those lipids that were more tightly bound to the ceramic matrix. These more tightly bound lipids could be extracted from the pottery using direct acidic or alkali extraction, but not chloroform/methanol extraction. It is possible that the more available lipids degraded so rapidly from this sherd because of the highly unsaturated and branched nature of many of the compounds present.

The residue from sherd 1 had not broken down as much as that from sherd 2—the residues obtained through chloroform/methanol extraction and acidic methanol extraction were more comparable. They are not identical, however, and deliver two different interpretations of the vessel contents. No isoprenoids were present in the chloroform/methanol extraction. Previous studies have noted that isoprenoid fatty acids are extracted more readily using direct acidic extraction (Correa-Ascencio and Evershed 2014). Interestingly, however, the chloroform/methanol extraction appeared to contain more unsaturated fatty acids than the acidic extracted residue. The relative amount of $C_{16:1}$ and $C_{18:1}$ is higher, as is the relative amount of $C_{16:0}$ compared to $C_{18:0}$, as can be seen in Table 7.3. Nonetheless, the residue obtained by chloroform/methanol extraction of sherd 1 cannot be interpreted as containing lipids from fish/shellfish, while the acidic hydrolysis extraction from the same sherd is interpreted in this way. The chloroform/methanol extraction did not contain any isoprenoids, nor did it contain $C_{22:1}$. Although the presence of 16- and 18-carbon monounsaturated fatty acids was higher than usual, it was not paired with any of the other indicators for fish or shellfish presence, and there are many plant lipids that contain similar abundances of monounsaturated fatty acids. Further, the ratio of $C_{18:0}$ / $C_{16:0}$ in the chloroform/methanol extraction, although less dominated by $C_{18:0}$ than the acidic methanol extraction, is still not consistent with the potential indicator for fish or shellfish. The chloroform/methanol extraction from sherd 1 did not contain any indicators for fish or shellfish and would have been interpreted as containing primarily plant lipids.

Clearly, acidic methanol extraction of residues is a much more effective way of extracting isoprenoid fatty acids and long-chain unsaturated fatty acids than chloroform/methanol extraction. In this case study, acidic methanol extraction was generally more successful than chloroform/methanol extraction. It extracted more lipids and allowed a more specific interpretation of vessel contents, including identifying the potential presence of fish and/or shellfish lipids in the residues. That interpretation was not possible using the chloroform/methanol extraction, because the isoprenoid fatty acids and longer-chain unsaturated fatty acids remained in the bound insoluble fraction of the residues with this extraction method.

Final Thoughts on Fish and Shellfish

The George Reeves case study shows two sherds whose residues are interpreted differently based on the extraction technique used. The reason for this is the elusive nature of fish and/or shellfish lipids. Although the results strike me as contradictory and unsatisfying, the identification of fish and shellfish lipids is often like this. Fish and shellfish lipids are so prone to hydrolysis, oxidation, and microbial breakdown that the extraction technique is probably crucial in their identification, as it probably is not when identifying most other foodstuffs. The lipids are likewise distinctive without being unique, resulting in the list of indicators already described, few of which are definitive.

If freshwater fish and shellfish identification in pottery residues is crucial (or even helpful) to your research questions, it's important to come to terms with these issues. You will need to check that your analyst uses either acidic methanol extraction or one of the other direct acid/alkali extraction methodologies that work best in removing fish and shellfish lipids from the residue matrix. You will need to communicate clearly with your analyst about your goals for fish and/or shellfish detection.

Then, after sending off the sherds to the lab, you will need to acknowledge the fact that the analyst's answer may not be definitive. In the best case, the residues are likely to be marked as containing fish and/or shellfish by two to three indicators and therefore interpreted as probably containing fish and/or shellfish contributions. In the worst case, it is possible for fish or shellfish lipids to degrade to the point where they cannot be identified in a residue, even using the most successful extraction technologies and the most sensitive instrument settings. This is particularly true for residues deriving from a mixture of resources, including freshwater fish and shellfish.

8

SAMPLE SIZE AND RESIDUE ANALYSIS

BROADLY, THERE ARE two ways of thinking about archaeological sampling: statistics-based and experienced-based. In statistics-based sampling, the project sample size is directly selected based on the type of questions asked, constructed testable hypotheses, and the statistics that will be used to test the hypotheses. This method is usually based on some form of significance testing, with sample size selected to obtain a desired range of significance. If you are interested in this sort of sampling, you're best advised to discuss the statistics for your study directly with a statistician or archaeologist with considerable statistical experience. Relatively few archaeologists use statistics in this way on a regular basis, however, and this type of sampling methodology is actually fairly rare in the field for residue studies and also for archaeological studies in general (Cowgill 2015).

More common in archaeology is an experience-based selection of sample size. As archaeologists gain experience, both in the field and during research, we become familiar with various archaeological questions and analyses and approximate acceptable sampling strategies for each. Often, of course, we're unable to submit the perfect sample size for an analysis due to funding issues or the destructive nature of some testing. Nonetheless, many archaeologists, myself included, select sample sizes for specialist analysis based on a body of internalized knowledge. The problem is that residue analysis is new enough that this internalized body of knowledge has not really developed, at least among most archaeologists. It's therefore unclear what an acceptable sampling size is to answer different types of archaeological questions. As with any other type of analysis, sample size affects what questions can be effectively asked and answered with residue analysis.

Given the breadth of issues that can be investigated in-depth by residue analysis, larger sample sizes will allow more interesting questions to be asked about the relationship between people and pottery. In fact, only through larger sample sizes can questions of greater archaeological interest and subtlety be investigated and the promise of organic residue analysis finally be fulfilled (Evershed 2008b). This chapter gives a basic framework for determining a reasonable number for different kinds of archaeological questions. It pays close attention to minimizing variables and the concept of functional groups in sampling. These concepts are illustrated by a detailed case study and two examples from the literature. Again, for carefully planned, statistically based studies, consult a statistician.

THE FISHING EXPEDITION (PRELIMINARY OR PILOT PROJECT)

One of the most common types of residue studies at the present time is what's often called a fishing expedition, or more formally a preliminary or pilot project. This is a small study that asks whether residues are present in pottery from a site or region, and what general sorts of interpretations can be made. Fishing expeditions are unfocused studies, with no specific archaeological questions, and often with no previous residue analysis performed on pottery from the site or area. The sample size is usually around ten to twenty sherds.

Even though there are no specific archaeological questions, and the research questions are broad, sampling is crucial in determining the results of fishing expeditions. Because sample size is small, variation in the pottery sampled may lead to results that are difficult to interpret in a meaningful way. However, if only one pottery type is sampled to produce more meaningful results, then clearly the project is only addressing that one type of pottery. So, even though there may be no formal archaeological research question in such a fishing expedition, there are implicit questions or limitations posed by the samples chosen. When putting together this sort of project, it makes sense to make these questions explicit, and then select the samples based on which results would be most interesting for a given site.

For example, an archaeologist has enough funding to analyze twenty sherds from a three-component site. Each component has many types of pottery, including what seem to be cooking vessels, decorated jars, and plates. In choosing the twenty sherds to sample, what should the excavator choose? There are many different permutations, each resulting in varying numbers of interpretive groups, numbers of sherds in each interpretive group, research question asked, and ability to answer the research questions. A short list of some of these permutations are shown in Table 8.1.

Table 8.1 demonstrates that in a study with a fixed number of sherds, as the number of different elements sampled (variables) increases, so does the number of potential interpretive groups and the number of research

TABLE 8.1. DIFFERENT WAYS TO SELECT TWENTY RESIDUE SAMPLES FOR PILOT PROJECT FROM A THREE-COMPONENT SITE WITH THREE MAJOR VESSEL FORMS

Pottery sampled	Time period sampled	# of interpretive groups/sherds in group	Research question	Confidence in answer to research question
One vessel form	One component	1/20	Did residues survive in cooking pots from Component A at the site, and what was cooked in them?	Fairly high
All vessel types	One component	3/7 or 6 (one group per vessel type)	Did residues survive in pottery from Component A at the site, what was processed in the vessels, and how did vessel use differ by type?	Fairly high on first, medium/low on others
One vessel form	All components	3/7 or 6 (one group per time period)	Did residues survive in cooking pots from the site, did their use/contents change through time, and what was cooked in them?	Fairly high on first, medium/low on others
All vessel forms	All components	9/2 or 3	Did residues survive in all pottery types from all time periods at the site, did their use/contents change through time, how did vessel use differ by type, and what was processed in the vessels?	Medium high on first, very low on others

questions. As the number of interpretive groups increases, however, the number of sherds in each group naturally decreases, as does the ability to interpret the data in a meaningful way. For example, the sampling strategy focusing on only cooking pots from component A at the site has a single interpretive group (cooking pots from component A) with twenty sherds. The analyst will be able to answer the two research questions with some confidence and will be able to discern at least some patterns in the data. As only cooking pots from component A have been sampled, however, the excavator has no idea about what is found in residues from the other components or from other vessel types. In this sampling strategy, breadth of findings is sacrificed to get interpretable results. This is often described as minimizing the variables, and it's important in maximizing the number of samples in a given interpretive group. In a pilot project, however, the archaeologist might be interested in knowing about the feasibility of residue analysis in all the vessel types and components at the site.

This is reasonable, and many pilot projects take a form much closer to the other rows in Table 8.1. Most common is probably the final row, where the excavator samples all different pottery types from all components. This results in nine separate interpretive groups (cooking pots from Component A, plates from Component B, bowls from Component C, and so on) of two or three samples each. With only two or three samples in each group, it will be nearly impossible to discern any pattern in the data at all, and any pattern that is detected could easily be coincidental based on the small number of samples in the group. For this sampling strategy, there are four implicit research questions in the study, and only one can be answered with much confidence. That is the question that functionally treats the whole study as an interpretive group of one—the question on whether residues survived at the site. The trade-off between the number of variables in a sampling strategy and the confidence in the results is crucial to all sampling and is present in all residue studies, whether a fishing expedition or otherwise. Because these small pilot studies are exploratory, however, often the research questions implied by the sample selection are not completely articulated, and the results (or their lack of interpretability) may come as a disappointment.

An archaeologist, when selecting sherds to sample for a pilot residue project, should be intentional about the types of sherds to choose and articulate what research questions are being posed by the sample selection. Once the questions are articulated, the archaeologist can then think about trade-offs in terms of breadth of the study vs. confidence in the results and determine their preference.

Ideally, fishing expeditions, especially broadly sampled ones with many interpretive groups, are a preliminary analysis—hence the more formal name. Once the archaeologist knows that residues are present and some preliminary results, they can consider the larger anthropological and archaeological issues at the site that residue analysis could address. At that point they can plan a more targeted type of project using the preliminary data to help frame the sampling for the larger study.

FUNCTIONAL SAMPLE SIZE AND MINIMIZING VARIABLES

Sampling for archaeological residue studies (and many other types of studies as well) is focused on the concepts of minimizing variables, as already discussed, and on the idea of functional sample size. The functional sample size is the number of sherds in a study that might reasonably be expected to contain the results the research design is focused on—it is basically the same as the interpretive group described in the example above. It follows, therefore, that the functional sample size varies depending on the archaeological and interpretive questions being asked. This sounds confusing, but becomes clearer in practice and is very helpful in both understanding residue results and choosing samples to analyze. For example, in the broadest sampling strategy of the 20-sherd preliminary project example above, the analyst could be fairly confident in the answer to the research question "did residues survive in pottery from the site," because the functional sample size for this question was twenty. When the question was focused on the difference between interpretive groups, the functional group size dropped to two or three.

Increasing variables will still cause problems in interpreting residues, even if the functional group size is large. In our example above, it seems likely that cooking pots and storage jars had more residues than plates—this is not always the case, but it is a common result of many residue studies (Charters et al. 1993a; Evershed et al. 1995). In the most extreme case, in which interpretable residues are found in 100% of the cooking pots and jars but in 0% of the plates, the analyst would determine with confidence that residues did survive in the pottery. Looking at the entire sample, however, the interpretable residue yield would be about 66%, which does not reflect the actual patterning in the data. When breaking the data down into the smaller groups of samples from each type, the analyst would see the patterns by vessel type and would be able to draw conclusions from them. With a sample size of six or seven for each grouping of vessel types, however, it is easy to miss more subtle patterns in the data.

Another important point is the role of prior expectations and assumptions when developing functional groups for sampling. From a statistical point of view, this is Bayesian sampling, in that expectations and assumptions feed into the functional group formation (Cowgill 2015). If the original assumptions are incorrect, then new concepts of functional groups can be formed. To use the fishing expedition example again, an informed residue analyst would expect plates to have fewer residues than cooking pots and jars. So when looking at twenty sherds sampled from the three vessel types and determining yield of interpretable residues from them, the analyst would probably mentally group the fourteen cooking pot and jar sherds into a functional sample including sherds that are more likely to contain residues. The six plate sherds would be in a functional sample made up of sherds that would be less likely to contain residues. If this assumption is correct, and residues were found in a much higher incidence in the jars and pots, then the research question about survival of residues in sherds from the site would be answered only with respect to that functional group. However, if the assumption is incorrect, and interpretable residues were found in all types of vessel sherds at a more-or-less equal frequency, the analyst would have to rethink. First, the analyst would answer the research question about residue survival with respect to all the sherds in the study—the functional group shifted, as the initial assumption was proved wrong. The analyst would then probably also recommend a further study focusing on plates. This sort of unexpected finding is often a good starting point for future research.

In essence, functional samples are a kind of mental map of expectations that may change while the analyst looks at the data. Every analyst uses some equivalent of the functional sample during data interpretation and analysis. If this concept is thought out ahead of time, however, it can make sampling easier and more effective and minimize surprise at results that really should not be unexpected. Further, it can allow clear recognition of when an initial assumption is wrong, which can open up new avenues for residue analysis.

Case Study: Maize along the Mississippi, Complex Sampling, and Functional Groups

As an example of the importance of keeping a grip on functional sample size to prevent surprises, I'll use the large study of maize residues along the Mississippi already touched on in chapter 6. This example also allows us to look at the many variables that can be present in large-scale residue studies.

The project succeeded in its first goal, which was to develop a way to identify maize in residues, as described in chapter 6. The second goal, and the one that really interested me as an archaeologist, was to determine whether thinner-walled, shell-tempered pottery was used to process maize more often than were thicker-walled, grit- or grog-tempered types. Was this type of pottery, typical of Mississippian cultures, developed in part due to its efficiency in cooking maize for long periods of time? Did form follow function, or was the pottery style driven largely by agency and cultural preference? Since causation could not be determined, I chose samples around the possibility of finding a correlation between maize processing and thinner, shell-tempered vessels.

There were potential issues of temporal correlation unrelated to my question of interest. Since thinner, more shell-tempered vessels become increasingly common on Mississippian sites, and since maize also came into increasingly common use on Mississippian sites through time, it seemed likely that I would find correlation between thinner, more shell-tempered vessels and maize use caused by temporal changes. In that case, any correlation I found between pottery stylistic change and maize adoption might be coincidental and based on time, rather than the technical advantages produced by thinner, shell-tempered pottery in cooking maize for longer periods, which is what I wanted to know.

In an attempt to avoid this issue, I worked with a number of really helpful and generous archaeologists from four different regions up and down the Mississippi (Nathanael Heller, John Kelly, T. R. Kidder, Jodie O'Gorman, Timothy R. Pauketat, Robert Salzer, and Mary Beth Trubitt). The idea was that as maize spread at slightly different rates at different points around the Mississippi, as did changes in pottery style, it might be possible to determine whether maize and pottery style changes always happened in tandem, or whether there was a lag in some locations. If the traits were always correlated, this would perhaps support the hypothesis that thinner, shell-tempered pots were used because they were more efficient for maize processing.

I wanted to compare sherds with and without maize, so it seemed safer to choose sherds from at least two time periods, which added another variable. Functionally, then, the project included four separate geographically based subprojects, three of which also had at least two time periods included. As can be seen in Table 8.2, some of these geographic projects had more variables and more sherds than others.

In addition, in the American Bottom and Lower Mississippi Valley I focused my sampling on the periods when maize entered the archaeological record. This would hopefully capture the development of thinner, shell-tempered pottery and its use in maize processing. If we regularly found thinner, shell-tempered pottery that had been used to process maize and regularly found thicker, non shell-tempered pottery that had not, it would also support the hypothesis that thinner, shell-tempered pots were preferentially used for maize processing, possibly because these sorts of pots would heat more efficiently.

This is a lot of variables. First, four different regions. Then, within each region, at least two different time periods. There was also variation in wall thickness and in temper, which included shell, grit, grog, and, in the American Bottom (one of my regions), limestone. Wall thickness is particularly problematic because it varies along a sliding scale, in part based on where in the vessel a sherd comes from (thicker at the bottom, usually). And finally, detectable maize vs. nonmaize. I thought that a sample size upward of 100 was manageable, given the time frame and the fact that residues would also undergo CSIA analysis. With 100+ samples, the number of variables present in the study was probably going to cut my final interpretive groups down to a small size. If a strong relationship was found between thin, shell-tempered pots and maize use in four different regions, however, it would be a really exciting result. The risk seemed worth taking, and a sample size of 129 sherds was selected.

Table 8.2. Sample Selection and Functional Groups for the Mississippi River Maize Detection Project

Geographic region	# of sherds / # interpretable residues	# sherds containing maize	Sites	Time periods
American Bottom	66/52 Mississippian and Emergent Mississippian	7	Mees-Nochta (11S1198)	AD 1000–1050
			Halliday (11S27)	AD 1050–1100
			Cahokia (11S34/3)	AD 950–1350
Illinois River Valley	12/12 Late Woodland	0	Audrey North	Late Woodland
			Koster East	Cal AD 600–890
			Evie	Late Woodland
			Hill Creek	Cal AD 1210–1410
			Apple Creek	Middle Woodland/ Late Woodland
				Cal AD 1280–1440
			Walsh	Cal AD 1225–1350
	5/4 Jersey Bluff	0	Starr Village	
	Total: 17/16			
Lower Mississippi Valley	4/4 Baytown	0	Reno Brake	Cal AD 200–650
			Bayou des Familles	Cal AD 1200–1400
	35/33 Coles Creek and Plaquemine	1	St. Gabriel Mound	
			Morgan Mounds	Cal AD 1000–1400
			Osceola	
			Emerson	Cal AD 950–1170
				Cal AD 700–1200
				Mid-fifteenth century
	Total: 39/37			
Wisconsin River Valley	7/6 Late Woodland	0	Gottschall Rockshelter	AD 600–950

Note: Study originally meant to look at the relationship between vessel temper and wall thickness vs. use in processing maize. Sherds were interpreted as containing maize lipids using the 2001 criteria for maize identification.

In this type of study, it's often helpful to first divide up the project geographically, as this is a controlling variable for cultural differences. Geographically, there are four interpretive groups in the project, shown in Table 8.2. Of these four geographic groups, the Illinois River valley had seventeen samples and the Wisconsin River valley had seven. After analysis it became apparent that only two of the geographical regions had enough samples to answer interesting archaeological questions within the region—the American Bottom, and the Lower Mississippi Valley.

Functional Sample Size

The temporal focus of my sherds in the American Bottom and Lower Mississippi Valley led to an unexpectedly small number of sherds used to process maize. This is particularly the case as this study also developed the maize detection technique. It was the first study using the technique, so we used a very cautious approach to maize identification. With more background research and follow-up analysis, as shown in chapter 6, more of the residues could safely be interpreted as containing maize than I originally realized. At the time, however, it seemed best to be cautious in my interpretations. Using the original criteria for maize detection, of the 129 samples in the study, only eight residues could be identified as containing maize, either through *n*-dotriacontanol or fatty acid analysis (see Table 8.2). Of these, seven were from the American Bottom, one from the Lower Mississippi Valley, and none from either of the other regions. This seems small, until we look more carefully at the functional sample size for maize in the study. In this case, that would be the number of sherds in the study that might reasonably be expected to contain maize. Table 8.2 also divides up the number of samples in functional group by time period. As we can see, the sample size of actual sherds was large, but the functional sample size of sherds that might have been expected to contain maize was somewhat smaller.

For example, it's unsurprising that no maize contribution was found in the Gottschall rockshelter sherds, as all the sherds were from Late Woodland levels of the site. The sample size of sherds likely to contain maize from the site, the functional sample size, was zero. It would have been amazing and informative to find compounds from maize in these sherds. But it was not very likely, and the excavator and residue analyst shouldn't expect to do so. At first, it seemed surprising to find no maize from the Illinois Valley sherds. Of the 17 sherds sampled from the region, however, 12 were from the Late Woodland period, in which maize was not commonly farmed in the region, and five from the Jersey Bluff, in which it was. The functional sample size of the region in terms of maize detection was five, rather than 17. It is therefore less surprising that no maize was found from this region than it originally appeared.

In the Lower Mississippi Valley, maize probably enters the diet during the Coles Creek period, around AD 1000 or so, and appears to be slowly added into the already rich diet there (Fritz and Kidder 1993; Kidder 2002; Listi 2011). The functional sample size from this region was 31, which is a reasonable sample size. However, the relatively slow adoption of maize in the region meant that it composed a noticeable, but not major, part of the diet in Coles Creek and even later Plaquemine Mississippian sites (Fritz and Kidder 1993). Despite the large functional sample size, the overall probability of finding maize in these sherds was lower than in a region in which maize made up the majority of the diet. As described, masking makes it likely that residues containing both maize lipids and lipids from C_3 fatty resources will tend to look more like the C_3 resources. The maize contribution will tend to be masked, unless it is the major contributor to the residue, or unless all the other resources processed in the pottery were lipid-poor. It was therefore disappointing to find maize-derived compounds in only one of the 31 sherds from the Coles Creek period or later. But, given what we know about diet in the Lower Mississippi Valley, it is only mildly surprising. The one residue that did contain a maize contribution from this region was HRL 208, a dramatic example of masking, discussed in detail in chapter 6.

The American Bottom region contained the largest sample size of any of the geographic regions, in terms of overall sherds and in functional sample size. In this area, sampling focused heavily on two sites that dated to the period when maize was being incorporated into the foodways of the people, as seen in Table 8.2. The Mees-Nochta site dated to the late Emergent Mississippian period in the floodplain (approximately AD 950–1050), where maize was present in about half of the features at the site (John Kelly, personal communication 2000). The Halliday site was one of the Richland Complex towns found in the blufftop outside the American Bottom during the early Mississippian period (ca. AD 1050–1150). It seems to have functioned as a farming center, and maize was an important crop at the site, although this was unclear when I originally took the samples (Alt 2002; Fritz 2019; Pauketat 2009). In addition, some sherds excavated from the palisade area of Cahokia were included. All 66 sherds sampled from the region might be expected to contain maize.

We therefore see that although there were 129 samples in the overall study, the functional sample size for maize overall was 106. When divided geographically, the functional sample sizes dropped to the numbers shown in Table 8.2. In short, although maize was discovered in only 6% of the total sherds in the study, it was found in 11% of the American Bottom samples. This is still a lower yield of maize contribution to residues than I expected at the time. The result now seems understandable, taking into account that the pottery from these sites was largely focused on early maize adoption in the area, masking issues, and the conservative interpretation of maize in fatty acids used at the time. To be sure of finding more maize, I should have sampled some pottery from later Mississippian periods in the American Bottom, when maize contributed more to the daily diet of the people. The focus on the hoped-for correlation between maize adoption and vessel style change led to a lower overall detection of maize in residues than expected. The result was a small number of residues with maize contributions, albeit from a large functional sample of sherds that might contain maize compounds.

In the end, the data suggested that, at least in my samples, maize residues did not seem to correlate meaningfully with thinner, shell-tempered pottery in the American Bottom and Lower Mississippi Valley, as shown in Figures 8.1 and 8.2. Even when applying the more recent interpretive criteria for maize detection in fatty acids, shown in chapter 6, there does not seem to be a clear correlation between temper type and maize use, as shown particularly in Figure 8.2.

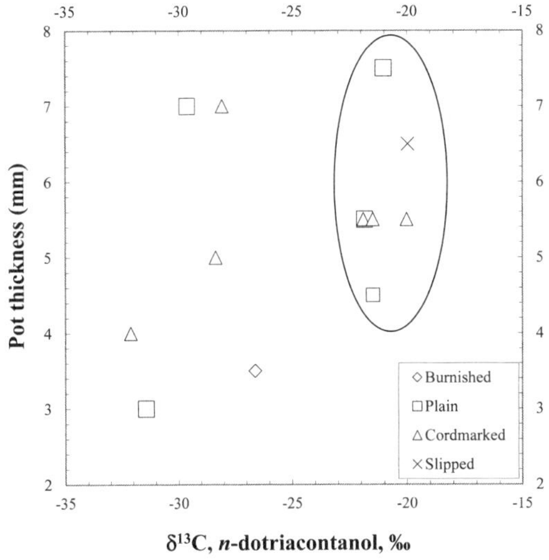

Figure 8.1. $\delta^{13}C$ of *n*-dotriacontanol plotted vs. pot thickness and tempering agent. Maize-containing residues are in circled area. (Eleanora A. Reber)

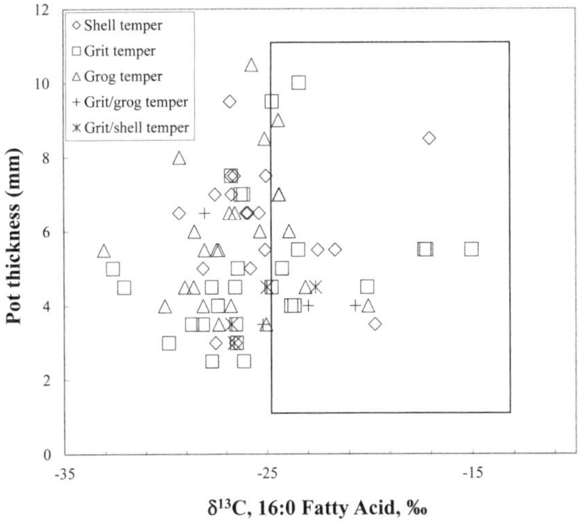

Figure 8.2 $\delta^{13}C$ of $C_{16:0}$, plotted vs. pot thickness and temper type. Residues with probable maize contribution are enclosed in the square area, using the more recent interpretive criteria for maize detection. This suggests a C_4 contribution to fatty acids more enriched than −25‰, as opposed to the criteria used at the time, which used more enriched than −21‰ to detect C_4 contribution to a fatty acid. (Eleanora A. Reber)

In sum, many problems in the sampling from this study derived from inattention to two guiding principles. I did not minimize the variables with enough ruthlessness, and at the time I had no clear concept of functional sample size. Although the sherds from Gottschall and the Illinois River valley provided some interesting results, and gave some interesting information about pottery use at the sites, the samples were not especially helpful in terms of one of the overarching goals of the project. They generally added variables and did not add many sherds to the functional sample of maize-containing sherds.

Coming up with a focused residue study involves real sacrifices. Interesting samples will have to be omitted, and it's important to stay focused on the central research goals of the study. In this case, I would have been better served by focusing entirely on sherds from the American Bottom and the Lower Mississippi Valley, which would have provided some geographic variation and allowed an increased sample size from these two areas. In addition, more samples from periods known to contain large amounts of maize should have been included, to counterbalance the sherds from periods that could not be expected to contain any maize.

NUMBER OF INTERPRETABLE RESIDUES AND SAMPLE SIZE

Functional sample size is often based on the number of interpretable residues in a study. For example, in the fishing expedition example above, plates would not be expected to contain much residue, and so they do not contribute to the functional sample for any question involving residue interpretation. If the research question includes uninterpretable residues as important data that is crucial to the central research plan, then samples without residue would be included in the functional sample. For example, a research question asking "how many of these vessels were used to store or process foodstuffs, and how many are empty or used to carry water" would use the presence of uninterpretable residues as a crucial data point. The functional sample and overall sample would be the same.

More likely, however, are questions about what people processed in pots, and often when. Focused studies tend to ask questions such as "were people processing plant foods (or dairy, or meat) in this type of pottery?" Analysts for these projects will need to consider the question of the yield of interpretable residues in the study. This is because in this sort of study, the uninterpretable residues do not contribute to the overall results and decrease the functional sample population.

In most of Europe and Great Britain, on average about half of all submitted sherds contain interpretable residues (Evershed 2008b), although the percentage is much higher from Scandinavia, where the cold weather seems to aid residue preservation (Isaksson and Hallgren 2012). The approximate average of interpretable residues in the southeastern United States, based on sherds submitted to my laboratory, is usually around 70–80%. In a study focused on interpretable residues, 40 sherds sampled with an 80% of interpretable residues will give a functional group of 32. With a 50% interpretability rate, the functional group will be only 20.

It's very difficult to predetermine the percentage of interpretable residues of pottery from an area without prior experience. Preservation issues in residues, as discussed, are not well understood but are clearly affected by soil chemistry at the sites, moisture contents, soil bacteria, rainfall levels, the original amounts of lipid absorbed within the vessel walls, and possibly the clay type in the pottery (Evershed et al. 2008a).

When determining your question of archaeological interest, one question that will help determine the number of samples to take is whether you are interested in looking at absence of residues (water, storage pots, etc.) or whether you are interested primarily in vessel contents. If vessel contents, then you should think about the probable percentage of interpretable residues in the study. Ideally, you can use an earlier preliminary project to act as a baseline for estimating yield of interpretable residues. Failing that, if the residue analyst has sampled in the area before, they might have an idea of the usual percentage of interpretable residues, which can be used as a guide. If not, it's perhaps best to start with 50% as a good estimate, but be ready to resample if your site has a low yield of interpretable residues.

Note: Why Are Average Interpretable Residue Yields Higher in North America Than Temperate Europe?

The higher yield of interpretable residues from North American than temperate European sites may be due to several different factors. Many of the sherds submitted for residue analysis in the southeastern or midwestern United States are from known or suspected cooking pots, while at least some of the vessels submitted from temperate Europe and Great Britain may have been water vessels and containers of dry contents. This is, of course not always the case—there is plenty of prestige noncooking ware present in the southeastern and midwestern United States. Since the analysis is destructive, however, these sorts of pots are submitted for analysis relatively rarely. The negative-painted pots submitted from Angel Mounds (chapter 7) are an unusual case. In addition, labs that can perform absorbed residue analysis are rarer in North America than in Great Britain and Europe, so that there are fewer submissions of samples overall in North America. It could be that yield of interpretable residues will revert to a global temperate mean when sampling increases in different parts of the world and includes a wider range of vessels.

Another possibility is in the resources commonly processed in vessels. Nuts were an important food resource for most North American Indian groups (Emerson et al. 2005; Fritz 2019, and many others). Most nuts, with the exception of chestnuts, are high in lipids, as discussed in chapter 5. A relatively small amount of nut processing in a vessel would lead to an interpretable residue. In addition, although game meat and most fish are relatively low in lipids, maize is relatively high in lipids compared with European staple grains, as are many of the Eastern Horticultural crop domesticates with the exception of squash, as discussed in chapter 6. As a result, the foods processed in North American vessels could have been higher in lipids overall than European ones, resulting in a higher overall interpretable residue yield.

Alternatively, it may be that clays used to produce pots in the southeastern and midwestern United States, or soils in these regions, are more conducive to residue preservation for some reason than are those in temperate Europe. This seems somewhat unlikely, but given the importance of ceramic structure and soil environment in residue preservation, the possibility cannot be completely discounted.

Finally, the yield percentages given in this section are taken from submissions to my lab of published and unpublished results and are compared with the University of Bristol's lab submissions of both published and unpublished results. It is possible that the difference is due to purely statistical variation, because my lab has processed far fewer samples than the Bristol lab has. Because so many results are not published, each lab tends to calculate its own average residue yields in-house.

MINIMIZING VARIABLES AND MAXIMIZING FUNCTIONAL SAMPLE SIZE: TWO EXAMPLES

Interpreting absorbed residue analysis is relatively simple if only one variable is chosen. For example, in the fishing expedition example, the sample that focused on a single vessel type from a single time period gave the most dependable results. That stated, there are so many different elements in pottery use that residues can address that it's hard to choose! Changes in vessel use through time, differences in use of different vessel forms, differences in use of different surface treatments or decoration elements, differences in use of pots from different locations . . . the list goes on and on. The crucial element is to keep track of how many questions you're asking at once. The more variables present in any project, the larger the required sample size. There are many interlinked variables in pottery, so in most cases it is probably impossible to decrease the variables to one. Too many variables, however, can lead to no real answers.

When doing a focused residue study, the key is to have a clear-cut research question with large functional sample sizes and to minimize all possible variables. This will involve careful conversations between the archaeologist and the residue analyst. To examine how this might play out, we will examine two focused studies that successfully minimized variables and maximized functional group size in two different ways.

Minimizing Variables

In 2018, a team at the University of York published a study focused on detecting starchy food processing in pottery from the Tianluoshan site in China (Shoda et al. 2018). Similarly to the project focusing on maize detection along the Mississippi, the paper was focused on two separate issues. The first was the detection of a potential biomarker for the processing of starchy plants. This would be helpful in detecting lipid-poor starchy grains that would be easily masked in a mixed residue. The second was the potential detection of rice in a residue—something of a holy grail in residue analysis, because rice is so low in lipids, and the lipids it does contain are typical of most plants (Gunstone et al. 1986). Instead of focusing on a wide range of regions, however, this study focused its sample population and minimized its variables rigorously.

All samples were taken from the Tianluoshan site, a well-known early Neolithic site with outstanding organic preservation due to waterlogging (Guoping 2013; Shoda et al. 2018). As such, it had clear evidence of early rice agriculture and unusually large amounts of visible pottery residues on potsherds. All samples were from the lower, Neolithic levels of this site to minimize regional and temporal variation. Absorbed residue samples were taken from a single pottery type, to minimize vessel variation. Visible residues seem to have been taken from a wider range of vessels, but all were clearly cooking vessels due to the presence of encrusted residues (Shoda et al. 2018). This study managed reasonable results with 20 absorbed residue samples and 36 visible residues, for a total sample size of 56. The rigorous minimization of variables, however, maximized the functional sample size, so that one of the functional sample groups included the visible residues, while the other included the absorbed residues. In the event, the study successfully identified levoglucosan as a potential biomarker for burnt cellulose or starch. Due to the number of starchy foraged foods in the area, however, it was impossible to distinguish rice from other starchy foods potentially processed in the vessels. The study did successfully identify starchy food processing in mixed residues, however, which is an exciting step forward in residue analysis (Shoda et al. 2018). By minimizing the variables and maximizing the size of the functional sample, the authors were able to determine clear patterns in the data with a relatively small sample size for this type of study.

Maximizing the Functional Sample

A similarly successful sample strategy, with a much wider range of variables and more expansive archaeological question, can be seen in an important study carried out by a team from the University of Bristol and several other universities investigating the relationship between animal domestication and dairying in southern Europe (Evershed et al. 2008a). This study was designed around testing Andrew Sherratt's secondary products revolution hypothesis, the idea that the initial domestication of cattle and other ruminants was for meat and hides, and that secondary products, such as milk, cheese, and yogurt, were developed after the initial domestication event (Sherratt 1981). A corollary to this hypothesis would be that if there was a secondary products revolution, dairying would probably develop somewhere outside of the initial area of domestication for the animals. If dairying was always a primary use of the animals, then it should show up earliest in the initial area of domestication. As a result, sampling for this study was very wide-ranging, including 23 sites spread across six regions (central/southeast Europe, northern Greece, northwest Anatolia, central Anatolia, southeast Anatolia, Levant) and time periods extending from around BCE 7000 to 3500. Having done this, the study then used CSIA to detect evidence for dairying, as described in chapter 6.

With so many variables necessary to investigate the potential advent of dairying across a wide range of geographic locations, the functional sample groups must be carefully monitored. In this case, the functional groups of interest are the sherds with residues from each region. As the study was explicitly looking for dairy or its absence, all sherds not containing interpretable residues did not contribute to this functional group. This study was complicated, therefore, by the unexpectedly low yield of residues from many of the regions—12%. As a result, the team needed to add more samples to the study until the number of sherds with residues reached a threshold of interpretability. The study eventually sampled about three times more sherds than they originally expected, with a total sample of 2,225 sherds (Evershed 2008b). Because of the relatively low yield of interpretable residues, however, the number of interpretable residues in the study was 255, with the smallest functional group, from southeast Anatolia, containing 13 interpretable residues (Evershed et al. 2008a).

The study found regional variation in dairying across the study region. Although dairying appeared earlier throughout the study region than expected for the initial secondary products revolution, the highest intensity in early dairying took place in northwest Anatolia, outside of the original area of cattle domestication. The data therefore seemed to support an early but variable adoption of dairying that may have taken place after the initial domestication and been variably important depending on each region's environment (Evershed et al. 2008a). The key importance of this study, however, lies in the disciplined use of continual sampling to guarantee at least a reasonable functional sample size for each region and time period. This allowed residue analysis to address a major interpretive question in archaeology for the first time. In addition, the data set, once begun, could then be included in other residue studies in the area, automatically increasing overall functional sample size for relevant studies in the regions of study (Whelton et al. 2018).

RULES OF THUMB: HOW MANY SAMPLES SHOULD I CHOOSE?

Determining a good sample size is in many ways a dialogue between the archaeologist and the analyst. Only the archaeologist can prioritize the research questions that can be addressed by pottery residue analysis. Only the analyst can determine how many sherds can be processed and interpreted in a given time frame, and for what cost. The two together can then decide on a good compromise between all the demands and trade-offs, so that an important research question is addressed in a timely and cost-effective manner. A key first step, however, is to determine general rules of thumb for more complicated archaeological problems.

The final question is therefore how many samples ought to be present in each functional group? This will again really depend on a trade-off between costs, desired results, and variables. However, each functional group should include at least 10 to 20 samples, with 10 being an absolute minimum that should be avoided if at all possible. After that minimum, more tends to be better up to a functional group size of perhaps 100? Given the relatively small number of large, multivariable residue analyses, a good upper limit for sample size has not yet really been reached.

For a preliminary study, therefore, in which the presence/absence of residues is of interest to the archaeologist, a good sample size is around 15 to 25 samples—10 is doable, but not optimal.

For other types of studies, focus on the functional sample groups that apply to your archaeological focus, and do your best to keep them above 15. This means that in a focused study on a question of archaeological importance, it's important to build time into the study for additional sampling, if necessary, or to work from a previous residue project that already sampled pottery in the area.

Preliminary projects or previous large-scale projects whose sampling overlapped your sites, time periods, or areas of interest can be used as a springboard for later studies, even if their primary area of research was not the same as yours. For example, the large-scale dairying project already discussed produced a large data set, whose interpretations can now be added to any study sampling in one of the regions or time periods included in the larger study (Whelton et al. 2018). Since full interpretations were done on all residues, whether dairy-containing or not, this data set can be included even in studies whose research questions do not focus on dairying.

Finding this kind of previous residue work in a region can sometimes be challenging. Publication of residue studies, especially preliminary projects with small sample sizes and unclear research questions, can be difficult. Most archaeometric journals will now accept only residue analyses that are of serious archaeological or methodological importance. This makes sense for the journals but means that most smaller and less dramatic residue analyses are published either in small regional journals or are not published at all. Some residue analysts have become discouraged from publishing such projects, as the rejection rate is high. The result is that a large portion of residue analyses, particularly preliminary projects but also larger projects with methodological flaws or undramatic conclusions, are submerged into what can be called the dark gray literature. Some of these analyses are present in site reports, often as appendixes, but others exist only in the files of the residue lab and the excavator. In this case, both the archaeologist and the residue analyst will need to ask around to determine whether anyone in the region of interest has completed any sort of unpublished residue project. If so, you can ask for a copy of the report, and if possible more of the raw analytical data, which will include information that can be used as a starting point for your project. In the future development of the field, it will be important for residue analysts to work together to form public databases that include such unpublished work. As the field progresses, interpretations are likely to change, and such databases will be useful both as starting points for further study and in refining previous interpretations.

If you are sampling in an area without a previous study, and your functional group is the number of interpretable residues, there should be some way to add more samples if necessary. Ideally, time for additional sampling should always be included in the original plan. Residue samples often have unexpected results that might change the direction of the original research plan, or that might require more investigation. It's better to plan for this ahead of time if possible.

9

THE BIG PICTURE

LIMITING FEATURES IN applying pottery residue analysis to archaeology include the relatively small number of labs doing the analysis and the relatively low throughput of samples that labs in many parts of the world can run. As well, as should be clear from the preceding chapters, the limiting factor in most laboratories is the number of trained residue analysts. As the analysis gains in importance and continues to become better known, it seems likely that more people will become interested in training as residue analysts. In time, residue analysis will become another commonly used technical analysis in archaeology, analogous to paleoethnobotany or faunal analysis.

At that point, the next step will be for residue analysis to integrate more tightly with archaeological questions, thus avoiding the "appendix problem" so common for technical analyses in archaeology. This is when technical analyses are relegated to the appendixes of a site report, separated from the conclusions and site interpretation. As residue analysis is so useful in examining the relationship between people and their pottery, this would be a waste of most of the potential value of residue analysis in archaeology. Instead, archaeologists should communicate with residue analysts to plan sensible sampling strategies that address archaeological questions of importance to the excavator. The first step in this plan is to understand what residue analysis can and cannot do. Hopefully, this book has helped with that step. The next step will be to think about archaeological questions that would be important to understanding the inhabitants of your site and that could be addressed by residue analysis. Then, you should talk these ideas over with a residue analyst to come up with a sampling strategy.

This chapter sums up the sorts of questions that residue analysis can answer at the present time. Methods development is always occurring, and new applications are probably in publication as you read this. But the overall framework for the sorts of questions that can be approached, and how, will probably remain helpful. Tips on how to communicate with your analyst so that your discussion is clear and comprehensible to you both are also included.

WHAT KINDS OF QUESTIONS TO ASK

This book has addressed some of the sorts of archaeological questions that can be asked in residue analysis. Time, space, and sanity prevented the inclusion of them all, though. Table 9.1 gives a summary of different approaches to archaeological questions using residue analysis. Each of these approaches can be applied differently, depending on the specific culture present at a particular archaeological site. For example, "determining whether different pottery forms were used differently" could include investigations of such artifacts as amphorae and potential perfume containers on Roman sites, but examine the difference between colonoware plates and bowls in early African diaspora sites in North America. It could include issues such as investigating the contents of different shapes of jars in the Levant, or determining whether stumpware at Cahokia was used to process ingredients or was a form of kitchen furniture. The extraction and analysis methodology will vary based on resources of particular interest. Analysts and archaeologists interested in caffeine, alcohol, or other psychoactives should consider the issues described in chapter 4 and seriously consider using targeted extraction and instrumental methods. Think of Table 9.1 as a sort of guideline that can be used to look at a wide range of archaeological questions.

In all cases, the relationship between people and pottery use is the central focus of residue analysis and the area in which archaeological questions can most helpfully concentrate. Residue analysis does not work well at addressing issues between people and food. Because so much food is processed outside of pottery, and because nonfood resources are processed in pottery, the interpretation of residue analysis focuses firmly on the relationship between people and ceramic vessels. This list is clearly not exhaustive, given the many possibilities inherent in residue analysis, but hopefully it will help spark a few ideas.

Another thing to keep in mind is that, as with most types of technical analysis, residue analysis tends to be more informative when paired with other lines of evidence. For example, residue analysis of dairying and milk production is most convincing when paired with faunal analysis from the sites involved (Evershed et al. 2008a). Detection of conifer resins in residues can usually only go to the family level botanically—to get a handle on species, it will be helpful to discuss the question with a paleobotanist. Residue, paleobotany, pollen, phytolyth, and starch analysis can all be used together

Table 9.1. Some Types of Research Questions That Can Be Answered by Pottery Residue Analysis, with Tips and Reminders

Research Question	Tips and Reminders
Determining whether vessel use changed through time on a site	Make sure functional sample size is large enough for each time period; minimize variables by focusing on single vessel types if possible.
Determining whether there was vessel specialization at a site	Often used to determine whether or not all similar vessels from a site were used to process the same contents. May want to consider spatial patterning on-site, as per Craig et al. 2015, or even household variation with enough samples and a large enough functional sample size.
Determining whether different pottery forms were used differently	Minimize variables as much as possible. Uninterpretable residues may be a helpful datapoint and therefore part of the functional sample.
Determining whether vessel with different styles/ surface treatments/tempers processed different contents	Minimize variables as much as possible. If residues from the site are often mixed, a larger sample size is advisable because differences between functional groups may be subtle (or absent).
Determining whether different sizes of pot processed different contents	Minimize variables by sampling similar vessel types that are in different size ranges. Keep an eye open for on-site spatial patterning in terms of both vessel contents and vessel size.
Processing and/or use of psychoactives (including caffeine and alcohol)	Will need to match extraction and instrument methodology to desired psychoactive, as in chapter 4, after conversation with the analyst. Easiest if there is already a hypothesis on what types of pottery/ container/artifacts were used for psychoactives.
Medicinal plants processed in pottery	Easiest to select samples hypothesized to have contained medicine, e.g., pipes, specially deposited vessels. For medicinal teas, it is believable that they might have been processed in cooking pots, resulting in a confusingly mixed and/or masked residue.
Detecting vessel sealing	Be familiar with common types of substances that might be used to seal vessels in your area, e.g., conifer trees, resinous nonconifers, phenolic resins, or bitumen/crude oil. Can be difficult to distinguish from production of sealants. Small, oxidized amounts of the biomarkers for the sealants may indicate sealing.

TABLE 9.1. CONTINUED

Research Question	Tips and Reminders
Detecting nonfood pottery use, such as resin production, maple sugar production, etc.	Easier if special vessels were used for these uses. If not, the resulting mixed residues can be hard to distinguish from vessel sealing. Usually, residues containing large amounts of the biomarkers for nonfood pottery use point to production.
Detecting the advent of new foodstuffs (maize, dairy products, etc.)	Easier if new foodstuff has some sort of biomarker or is isotopically distinguishable. If not, a larger-than-usual sample size with time for additional sampling is crucial.
Discerning regional or temporal patterns of adoption of new foodstuffs	Same as above
Discerning regional or temporal variation in use of pottery	Need to keep a close eye on functional group size, because the number of variables will be very high. Should have a clear hypothesis about the variation in pottery use, why it might happen, and how it might be tested, and then sample focusing on this hypothesis. Variations in pottery use are also likely to be subtle, so a larger-than-usual sample size with time for additional sampling will be very important.
Discerning regional or temporal variation in use of a detectable resource	Same as above, except that the automatic focus on a particular resource tends to make hypothesis construction easier. Still very important to keep functional sample size up, and allow time for additional sampling as necessary.
Determining vessel use at feasts/ritual events	Particularly good for obvious feasting/ritual deposits, but it is helpful to have a comparative baseline of nearby residues from pottery that was not used at feasts or ritual events. This adds another functional group and doubles the initial sample size, unless previous residue analysis has already taken place in the area.

in projects that are focusing on the detection of a particular resource, such as maize or some other plant. Many identifications of wine in the archaeological record paired residue analysis with archaeological evidence for grape processing (Badler et al. 1990; McGovern et al. 2017).

This is where you, the archaeologist, come in. As leader of the excavation team, you are the only person who can harmonize the research goals of the different technical analyses with your own excavation goals and strategies. The residue analyst will not be sufficiently familiar with the archaeology of your area to independently construct an archaeological hypothesis, find collaborators in faunal and paleoethnobotanical analysis (or whatever is most useful), and make the whole thing fit in with the sherds excavated. That means that communication between the archaeologist and the residue analyst (and all the other technical analysts) is foundational to the success of your residue project.

HOW TO TALK TO YOUR RESIDUE ANALYST

This communication sounds easy and a bit irritatingly holistic. It can actually be surprisingly difficult in some cases, because chemists and archaeologists have different technical vocabularies. This problem will not be as serious when working with an established residue analyst who has worked with archaeologists regularly. In the case of the most experienced analysts, they usually have picked up enough of the archaeological worldview and vocabulary that there will be no communication problem at all.

When dealing with a less experienced residue analyst or with someone in training, real problems can arise. First, be aware that when a lipid analyst says "clean," they are specifically referring to "clean of all lipids and contaminating compounds." Many things described as "clean" by an archaeologist (or anyone who isn't a lipid analyst) are not clean to a lipid analyst. Soap, for example, especially with added fragrances, is not clean in the residue sense. Sterile swabs are also probably not clean, because the swab is attached to the stick by some contaminating adhesive, and the sterilizing process removes microorganisms but not lipids. Tap water is definitely not clean in this sense. If the analyst asks for the samples in clean containers, and they are sent in clean plastic bags, it's not ideal. This is because of the contamination issues caused by plasticizers, discussed in chapter 3. The analyst wanted the sherds placed in solvent-washed tin foil or glass, ideally, or plain tin foil, and then the tin-foil wrapped sherds could be placed in the plastic bags. Such communication issues can cause real problems if not recognized and addressed early in the project (Mazow et al. 2014).

Clarifying these issues is particularly important if the archaeologist needs to do some of the residue sampling. Usually, sherds can simply be mailed to the residue lab and all sampling can take place there. Depending on archaeological permitting, national laws, institutional policy, or practicality, however, the excavator may need to flake off a visible residue or powder some of the sherd away from the residue lab. If this is the case, be certain you understand what the analyst means when they tell you the process. Ask for written procedures (it's easy to skip a step without a written protocol, and most labs have them on file). Make sure that when you are supposed to use solvents, you are using the correct solvent and grade for the type of analysis. Residue work generally requires extra-clean scientific-grade solvents, rather than the grades used in standard undergraduate teaching labs, for example. When the procedures ask for a “clean razor blade” or other tool, be aware that they expect the tool to be washed with high-grade solvents or burned at 400°C for a particular length of time.

If you’re solvent-washing, it’s wise to check on the ventilation in your sampling area and then ask the analyst if acetone can be substituted for the standard mixture. Many solvent mixtures used in residue analysis are toxic in closed spaces and should not be used without a fume hood. If you’re unsure which solvent to use, be sure to ask. The use of an incorrect solvent is surprisingly hard to correct later in analysis and will at least double sampling time. The use of a contaminated solvent results in the loss of the sample and cannot be corrected. In general, this sort of conversation is easier to have in a direct conversation (in-person, telephone, video call) rather than by email, because there are so many details to go over.

Aside from these more technical matters, you will also want to clarify practical issues with the residue analyst. First, how many samples can their lab comfortably take, and how long is the turnaround time? What does the analysis cost? Communicate any time constraints to the analyst up front to help with project planning. Be sure that some form of report is included with the analysis—it usually is, but it never hurts to check. At present, all residue labs with experience in the field are based in academic departments. Turnaround time varies from place to place, but most will be on an academic sort of time schedule with a longer turnaround time than might be standard for commercial firms or for a CRM firm. CRM firms can and do work with residue analysts, but it tends to be easier on everyone if the sherds are sampled and mailed toward the early end of the project timeline, allowing enough time for a more thorough analysis and interpretation period.

A Note on the Cost of Residue Analysis

It’s actually surprisingly difficult to come up with a standard price for residue analysis. For one thing, there is no commercial lab regularly performing this analysis, so there’s no baseline price for nonacademic research. Each academic lab has its own pricing model that includes different parts of the extraction and analysis process, depending on university policy, laboratory setup, and pricing structure for instrument use.

In my lab at the time of writing, for example, for a standard untargeted residue analysis that does not include CSIA, I charge around $65/sample. This price includes laboratory consumables, all instrument *analysis, and the production of a report. It does not cover the cost of my labor, however; I view residue analysis as part of my research responsibilities at UNCW. Because of this, my turnaround time is normally about a year. The analysis itself does not take that long, but scheduling the blocks of time for the analysis around my teaching and service responsibilities can only easily be done in about a year's time. So the price does not include labor but has a slow turnaround. For projects requiring shorter turnaround, I often double the price, which allows me to hire an assistant. Other labs might include labor costs or might charge per injection on the instrument. In a few labs, there might be separate charges for extraction, analysis, interpretation, and report writing.*

Additional extraction methods and instrumental analyses will increase the price of analysis but will also be added in differently depending on the lab in question. Double extraction for targeted analysis, as well as untargeted analysis, will increase the price variably, depending on whether the lab charges for labor and consumables or only for consumables. Additional instrumental analysis, such as HPLC/MS or CSIA, are often negotiated differently at each residue lab. Some already own or control a HPLC/MS or a GC-IRMS, and so scheduling and pricing are organized in-house. In my lab, however, I use instruments controlled by other faculty around UNCW, and the collaborating faculty member determines the price. In sum, there is really no one standard pricing method for residue analysis at the present time, and it is important to ask your residue analyst what pricing model they use.

Another issue to consider is that prices sometimes change. In my lab, there has been a slow but steady upward pressure on analysis prices caused by increases in the price of many laboratory consumables, and especially on the price of helium. The price of analysis at my lab has generally changed every two to three years, based largely on the costs of helium and instrument maintenance.

Then, on the most abstract but hard to communicate level, you should discuss your research question with the analyst. Depending on their level of experience, they might be more or less helpful when discussing archaeological variables. All should have some grasp of the residue analysis variables, however. These include optimal collection strategies for you to minimize contamination in the field, or types of research questions that are likely to be difficult to answer without a large sample size. If your research question focuses on a resource without a clear biomarker or isotopic indicator (such as nuts, for example), the residue analyst should be able to let you know that clear conclusions about this research question might be challenging. If your analyst is an academic residue analyst or graduate student, they may consider adding a method development section to the project, to try to develop ways to detect this resource, through either biomarkers or a series of indicators. Be aware, however, that method developments are not always successful. You and your analyst should come up with a backup plan together, in case the resource of particular interest does not have a biomarker or reliable indicator. Although some helpful information is included in this book, it is impossible to cover every known resource in residue analysis. Research in the field, moreover, makes it likely that new biomarkers and indicators will be developed in the near future. The residue analyst will be conversant with the recent literature and should be able to answer your specific questions about resources and the feasibility of your research questions.

Your residue analyst should also be able to tell you if one of your research questions can be answered by a residue analyst, but not by their particular lab. For example, not every residue lab has the GC-IRMS necessary for CSIA detection of maize, dairying, marine species, or any of the other resources discussed in chapter 6. If this is integral to your research question, you need to ask whether the analyst has access to this sort of equipment. The same is true for targeted methodologies and for HPLC/MS. If the analyst does not have access to equipment needed for your research question, they can probably give you an idea of who might be able to do the analysis for you.

BUILDING A LARGER SAMPLE SIZE AND OPTIMIZING FUNDING

Conversations between archaeologists and residue analysts are often fruitful of research projects that answer large archaeological questions, as summarized in Table 9.1. These require larger sample sizes than the standard pilot project. Depending on the individual circumstances of the lab and archaeologist, funding of some sort may be needed to complete larger, more intricate studies. The archaeologist may not have the budget for residue analysis that would permit the 200-sherd sample probably necessary to answer their research question. The residue lab may not have the personnel to run 200 samples in a reasonable time (less than a year). A CRM firm or graduate student might not have the six months needed to get the samples extracted and analyzed before results can arrive. Each of these problems may require different solutions. In all these cases, however, it would be worthwhile to seek some sort of outside funding. Lab personnel can be trained and increase a lab's throughput surprisingly rapidly, if there are funds on hand to pay them for a particular project. Some pieces of equipment can add to a lab's speed of analysis to a surprising degree. It is worth writing grants, in many cases, to obtain this outside funding.

Grants may be difficult to find, however, and creativity may be needed to complete a desired project. For example, an academic archaeologist might be able to break a large residue project into multiple, smaller projects that can later be combined. A study with three different regional components can be run in three consecutive years as smaller projects and then combined into its final form at the end. The result is three smaller but still useful projects that might be easier to fund through the normal excavation budget.

CRM firms can map out a long-term strategy and regularly submit smaller projects with carefully chosen samples that can be added into a larger data set that can be applied moving forward. Using this plan, even very small sample sizes of fewer than 10 sherds can be useful. The results can be slowly added into a regional and temporal database that will be increasingly valuable in answering larger questions with residue analysis. Such a strategy allows faster results paired with the sort of regional data set that is particularly valuable to future archaeologists asking related questions.

All these solutions should be discussed with the residue analyst ahead of time. Although mid-course corrections will likely be necessary, some sort of clear, feasible plan at the beginning of the process will be valuable in mapping out strategies as the project continues. Further, breaking projects up into smaller, more focused groups has the additional benefit of allowing tweaks to sampling strategies to reflect any surprising results that may appear in earlier studies.

Neither archaeologists nor residue analysts should shy away from using residue analysis from other archaeologists or analysts as a starting point. Even small studies whose results might be questionable can serve as a starting point for future research. Such studies can be viewed as preliminary projects that have already been seeded in the literature. Some analysis of similar residues might be necessary to test the conclusions reached by the other residue analysts and archaeologists, but all the work and interpretation of a preliminary project has already been completed. Again, residue analysis is surprisingly similar to archaeology. Residue analysis is destructive, and valuable data should not go to waste because methodologies and interpretation have improved through time. This is very like using data from older archaeological excavations when possible, even if excavation methodologies have improved in the previous 100 years.

EXCAVATING YOUR RESIDUES

Residues can be complex, difficult to interpret, and affected by a range of factors, from human behavior through preservation issues and postdepositional contamination. The same, of course, can be said about most archaeological sites. Archaeologists are therefore mentally well equipped to deal with the uncertainties of residue interpretation. The interpretation of pottery residues, although often ambiguous and including a variety of compounds from different uses, is functionally similar to the interpretation of most archaeological data.

Finally, I'd like you to take away from this book a feeling of excitement for the potential benefits of residue analysis in archaeology. Each residue is like an individual archaeological feature. With residue analysis, every archaeologist with pottery suddenly has hundreds and thousands of new features to explore! Sampling strategies are necessary to make this exploration time- and cost-effective. But the application of residue analysis to archaeological pottery, despite its roots in the 1990s, is still a new field for investigation. An explosion of new data and approaches lies ahead, limited largely by the training of residue analysts and laboratory capacity.

GLOSSARY

Some of these terms do not often appear in the literature but have been included so that you can communicate more effectively with an analyst.

A solvent: in HPLC/MS, this is the first solvent in the mobile phase. The method will begin with 100% A solvent flowing through the column.

Aliquot: chemical term for a somewhat undefined fraction of the whole. This is generally used for preparatory processes that use small quantities. The analyst may be taking ten drops from a twenty-drop sample, so it's impossible to give a precise metric amount, or even an exact fraction, of the sample taken.

APHs: ω,o alkylphenyl fatty acids, a stable molecule formed by pyrolysis of polyunsaturated fatty acids that acts as a biomarker for the presence of both highly unsaturated fatty acids and high heat.

B solvent: in HPLC/MS, the second and final solvent in the mobile phase. The method will end with 100% B solvent flowing through the column.

Baseline: "floor" of a GC or LC. Baselines should be flat and level, as discussed in chapter 2.

Blank run: when an instrument is run with no injection of any kind to look for ghost peaks. Blank runs should be completely empty.

Bound insoluble fraction: portion of a residue that is not removed from its matrix or substrate during extraction, presumably because it is chemically or physically bonded to the ceramic or charcoal. Compounds in this fraction are often more branched or carboxylated than those that are easily removed from the matrix. Alkaline and acidic extraction methods are more successful at removing compounds from their matrix, and therefore there is less bound insoluble fraction remaining.

CAM: crassulacean acid metabolism. This type of photosynthesis occurs in succulent plants and produces stable carbon isotope values that may mimic those from C_4 photosynthesis. In this type of photosynthesis, CO_2 is concentrated in the plant tissues during the nighttime, allowing photosynthesis to occur throughout the day while minimizing evaporation from the plant.

Capillary column: chromatographic column that is very thin (less than 0.5 μm internal diameter, usually) and with a siloxane-based stationary phase bonded to the inside of the thin silica tube. These columns are very long, flexible, and thin and provide better separation for compounds than the previous, packed columns. These were shorter, thicker, and packed with different forms of activated silica. All modern chromatographic methods use capillary columns; packed columns are now reserved for different types of sample preparation.

Carrier gas: in GC analysis, the mobile phase (chapter 2) that carries the sample through the column is the carrier gas. This can be several different gases, but helium (He) and hydrogen (H_2) produce the best separations. Helium is usually preferred due to its unreactive nature and because hydrogen may explode with the presence of a spark in areas with insufficient ventilation.

Chaîne opératoire: residue formation from vessel firing through burial, including contributions to residues from pottery firing, sealing, and mixing. Publications discussing residue formation over the lifetime of a vessel often use this term.

Coelution: when two peaks arrive at the detector from a chromatographic column at the same, or almost the same time. Two separate compounds coeluting can cause a misidentification of the mass spectrum, as it contains two different compounds, but there will usually be a distortion to the peak shape suggesting that a coelution of two compounds is present. Coelution can also be used to make a positive identification, as in an external spike.

Column type: each capillary chromatographic column is described by its length, thickness, and the coating on the inside of the tube, which is the stationary phase discussed in chapter 2. The longer and thinner the column, the better the separation, but the longer the run. The silica-based stationary phase of the column may have one of many different possible polarities, chosen to optimize separation of the sorts of compounds most processed in a lab. The overall column type will be chosen to obtain the best separation of most compounds in the fastest reasonable time. In most cases, the column will already be chosen by the lab director. New types of analysis for a lab, however, sometimes require new types of columns. Changing columns involves the complete replacement of all ferrules and often requires at least a day to restabilize the system.

CSIA: compound-specific isotope analysis. This technology allows stable isotope analysis of separated compounds within a complex mixture, such as a residue.

DAGs: diacylglycerols. Compounds with two fatty acids attached to a glycerol backbone, these compounds commonly result from the loss of one fatty acid from a triacylglycerol by hydrolysis.

Deconvolution: in GC/MS, a computer system or method that allows the determination and identification of coeluting peaks. The system follows the abundance of important molecular fragments, allowing coeluting compounds to be separated and identified.

Derivatize: adding functional groups to reactive portions of compounds in a gas chromatographic sample to make the compounds more volatile. This allows the derivatized compounds to be analyzed more quickly and completely, improves compound quantification, and may also allow the determination of a compound's chemical structure. Methyl esters and trimethylsilyl (TMS) esters are common derivatized compounds produced in archaeological residue analysis.

Derivatizing agent: compound added to a residue sample to derivatize compounds within it. These derivatizing agents often add modern carbons to an archaeological compound, which needs to be taken into account during CSIA analysis.

DEHA: diethylhexyl adipate, a common plasticizer.

Diagenesis: physical and chemical changes that take place in an artifact during burial.

DOSS (dioctyl sodium sulfosuccinate): a surfactant used in the dispersants applied to the Deepwater Horizon oil spill.

EIC (extracted ion chromatogram): a type of chromatogram that reflects the presence of one, or a small number of, molecular ions out of the range of scanned molecular weights. It allows easier detection of less abundant targeted compounds within a mixture.

Electron ionization (EI): most common way of producing mass fragments in GC/MS analysis of residues. The compounds are fragmented by bombarding them with electrons produced by a heated wire filament. The mass spectrometer then detects the mass-to-charge ratio of the fragments. Voltage running through the filament may be increased or decreased to control the number of electrons colliding with the compounds of interest.

Electrospray ionization: method of introducing the mobile phase in HPLC into the MS. The liquid mobile phase is sprayed into a fine mist using a nebulizer into a chamber filled with dry nitrogen, which evaporates much of the solvent. The separated compounds then flow into the MS, which is kept at high vacuum.

Elution: when a compound emerges from the end of a chromatographic column.

External calibration curve: known amounts of known compounds are run in different concentrations to permit an understanding of compound linearity for the known compounds in a method. This permits true quantification of the compounds in the external calibration, as opposed to semiquantification.

External spike: small amount of a known compound added to an archaeological sample to make an absolute confirmation of an identification. This is used especially when the mass spectrum is unclear, as in SIM analysis, or is not used in a given method. If the known compound coelutes perfectly with the unknown, they are the same, and identification is confirmed.

FAME fraction: fatty acid methyl ester fraction. When a residue is extracted using chloroform/methanol, a portion of the total lipid can be saponified and separated to produce a fraction containing all the fatty acids in the residue from all acyl lipids. It is the optimal fraction for looking at fatty acid abundances. Methyl esters are often preferred because it is better for CSIA, as only one modern carbon is added to each fatty acid during derivatization. The alkaline methanol extraction also produces this type of fraction.

Ferrules: soft, shaped plugs that are present in all or most connections in the GC or HPLC systems. Connections in GC and LC systems must be absolutely air- or watertight. Any leaks cause loss of sample and contamination from air (in the case of GC), but there are many connections in GC/MS and HPLC/MS systems. When tightened, a ferrule deforms around a connection joint, maintaining a tight seal. Ferrule materials must be carefully chosen so that they do not contaminate MS results. Because they deform with tightening, ferrules are almost never reusable and must be replaced every time a connection is taken apart.

FRE: freshwater reservoir effect, an issue in radiocarbon dating in which old carbon from limestone bedrock is incorporated within fish or shellfish tissues, producing dates that are older than the age of the fish or shellfish.

Freezer blank: in experimental residue studies, the sherd with experimentally produced pottery residue that is kept in a freezer while other identical residue-containing sherds undergo some form of simulated (or real) burial. This allows the analyst to determine the amount of oxidation and hydrolysis taking place in the experimental residues due to environmental or depositional conditions.

FTIR: Fourier transform infrared, a type of spectrometry that uses infrared wavelengths to produce an absorption spectrum of atomic bonds in a compound, allowing its identification. Most effective on single compounds.

GC: gas chromatography, a technique that separates complex mixtures of compounds based on their polarity and boiling point. This technique works best on compounds that evaporate below 300°C or that can be manipulated through derivatization so that they evaporate below 300°C.

GC-IRMS: gas chromatography isotope ratio mass spectrometry. The instrument used to obtain CSIA data. It separates the compounds in a mixture using gas chromatography, combusts each separated compound in a combustion chamber, and then measures the stable compound isotope ratio of each separated compound.

GC/MS: gas chromatography/mass spectrometry, an instrument that separates complex mixtures of volatile compounds based on their polarity and boiling point, and then identifies each separated compound by fragmenting the separated compounds into molecular fragments, whose mass is then measured. This is commonly used on lipid residues in archaeology, especially untargeted analysis.

Ghost peaks: peaks on a chromatograph that do not originate from the sample. They are usually detected by running sample blanks or instrument blanks—runs with only clean solvent or no solvent injected. Ghost peaks may derive from contaminated solvent but are usually from an overconcentrated on-column or splitless injection that did not completely volatilize or dissolve all the compounds in the sample. The system must be cleaned to ensure that all ghost peaks are eliminated before analysis continues.

HPLC: high-performance liquid chromatography/mass spectrometry, an instrument that separates mixtures of compounds based on their solubility in one of two solvents and then identifies each separated compound by fragmenting the separated compounds into molecular fragments, whose mass is then measured. It is particularly useful for targeted analysis, and for analysis of compounds whose boiling point is above 300°C.

Hydrolysis: chemical reaction that usually takes place in the presence of water. It is common in residues after burial and occurs when a water molecule reacts with bonds, often oxygen bonds, resulting in a compound breaking into smaller pieces. Hydrolysis is responsible for the breakdown of triacyl lipids into fatty acids, monoacyl glycerols, and diacyl glycerols, for example.

Injector temperature: in a GC with split/splitless injection, the injector port will be heated to a temperature that should volatilize all or most of the injected sample immediately on injection. The evaporated compounds then enter the column, where they usually condense on the front of the column until it is heated to the compound's evaporation point. This ensures that all compounds on the column will evaporate during the method and optimizes their separation. Injector temperature is usually set as part of method development, but if a new compound of interest is analyzed, the analyst should double-check that the compound evaporates at the injector temperature or below. Otherwise, the compound will tend to stay in the injector and will not be analyzed.

Interface: in a GC/MS or HPLC/MS system, the connection between the chromatograph and the mass spectrometer.

Interface temperature: temperature of the connection between the GC and MS on a GC/MS system. If the interface is not heated, compounds will condense in it, resulting in poor separation and ghost peaks. The interface temperature should be high enough to maintain the evaporation of all compounds of interest while minimizing thermal wear and tear on the interface connection. Higher interface temperatures are associated with shorter column life and connection problems at the interface.

Internal standard: known amount of a known compound added to a residue sample during sampling. This is used to semiquantify the amounts of archaeological lipids during the run. It is essential that the internal standard is a compound unlikely to exist in the archaeological sample, and it should ideally elute toward the middle of a chromatographic run.

IRMS: isotope ratio mass spectrometer, a sensitive mass spectrometer that can distinguish the weights of different isotopes from the same elements. These spectrometers can be used to measure isotopes of many different elements.

LC: liquid chromatography, a technique that separates mixtures of compounds based on their solubility in one of two solvents. It is especially useful for targeted analysis and for separation of compounds whose boiling point is above 300°C.

Linearity: during compound quantification, the quality in which the area under a peak is proportional to the known amount of a compound present in a sample. Linearity is best usually in the middle portions of a GC run, with the relationship between peak area and compound amount varying slightly at higher and lower temperatures, and at very small and large amounts of a given compound. This is why the internal standard should ideally elute toward the middle of an instrument run. Each compound in a residue will have slightly different linearity based on its elution time, elution temperature, and the unique physicochemical properties of the compound. The better the linearity of a compound, the more accurate its quantification.

MAGs: monoacylglycerols. Compounds with one fatty acid attached to a glycerol backbone, these compounds commonly result from the loss of two fatty acids from a triacylglycerol by hydrolysis.

Maillard reaction: chemical reaction that causes browning in most cooked foods. It is caused by a series of complex reactions between proteins and sugars.

Mass-to-charge ratio (m/z): fragments from a mass spectrometer are identified through their mass-to-charge ratio. The charge is generally 1, so the m/z ratio is the same as the molecular weight of a given fragment.

Methyl esters: derivatized compound in which a methyl group replaces a hydrogen in hydroxyl groups, so that –OH becomes $-OCH_3$. This makes the compounds more volatile and produces more sensitive, quantifiable analysis.

It is the preferred derivatization for fatty acids, especially those undergoing CSIA, because the procedure adds only one modern carbon to the archaeological compound.

Microbial degradation: breakdown of compounds in a residue through their digestion and processing by bacteria. Microbial degradation tends to result in more odd-chain fatty acids and alkanols and even-chain alkanes. The exact pathways by which compounds are transformed by microbes vary depending on the microbe and details of soil chemistry. This is a major contributor to residue transformation during burial.

MS: mass spectrometry, a technique where a molecule is fragmented by either electronic or chemical means, and the mass of each resulting fragment is measured. As the molecule fragments at consistent weak points, this technique is commonly used to identify unknown compounds.

MS-MS: type of instrumentation in which compounds after chromatographic separation (either GC or HPLC) are introduced to two successive mass spectrometers. The first MS fragments the compound as usual. The second MS further fragments the elements from the first MS, allowing the analyst to examine molecular fragments of molecular fragments. This allows more precise identification of compound chemical structures and much more sensitive analysis of target compounds. It does not obtain a standard mass spectrum for overall identification. Instead, the analyst focuses on the fragmentation patterns of interest for different compounds. This type of instrumentation is excellent for targeted analysis.

On-column injector: type of injector in which the complete sample is injected directly onto the chromatographic column. This ensures that no sample is lost during injection and is optimal for samples with very small amounts of the compounds of interest. Using this type of injection, however, compounds that do not volatilize or dissolve on the column will be introduced to the front of the column, often causing buildup that can cause ghost peaks and shorten column life. To prevent this problem, a retention gap is often used. When injecting samples with highly concentrated compounds with this type of injector, the samples must be diluted so that they don't overwhelm the column and the detector.

Oxidation: chemical reaction in the presence of oxygen. In organic residues, oxygen often reacts with double bonds. This reaction is responsible for the loss of unsaturated and polyunsaturated fatty acids from residues. Oxidation is complex and results in a range of products.

PAHs: polyaromatic hydrocarbons, a stable compound type that is commonly formed at very high temperatures and/or pressures. It is a common biomarker for the presence of soot and fire and also occurs in oil.

Peak fronting or front-loading: in chromatographic peaks, peaks with slopes on the front edge and very sharp rear edges are described as fronting or front-loading. This shape is typical of a sample in which too much of a given compound has been introduced to the column. Linearity will be compromised, and the analyst will need to run a blank to ensure that all the sample has been removed from the column. Peak fronting sometimes results in ghost peaks from the fronted compounds in succeeding runs.

Peak tailing: in chromatographic peaks, peaks with a sharp front edge and a long sloped rear edge are described as tailing. Peak tailing may derive from an old column or a compound that is not evaporating (in GC) or dissolving (in LC) optimally, due to problems with the stationary phase, mobile phase, or instrument method. Some tailing is acceptable, but too much affects linearity.

Pegged peaks: in CSIA analysis, these are peaks that are so large that they overwhelmed the column and mass spectrometer. Pegged peaks are mesa-shaped, with sharp front and rear edges, but a completely flat top. These compounds have overwhelmed the mass spectrometer to the point that the isotopic analysis is completely unreliable, and the sample will need to be diluted and rerun. The analyst will also have to check for ghost peaks.

POE: polyoxyethylene, as in POE fatty acid esters. They are commonly used in emulsifiers and surfactants.

Polyunsaturated compounds: compounds with more than one double bond present. The more double bonds, the higher degree of unsaturation. As double bonds are prone to oxidation, polyunsaturated compounds are often lost during use and/or burial and are rare in archaeological residues.

Procedure blank: blank sample that is prepared in parallel with all archaeological samples to detect contamination from laboratory procedures and materials. The procedure blank undergoes every preparatory process in the lab along with the archaeological samples and should be clean during analysis except for an internal standard, if added.

Pyrolysis: relating to fire. Pyrolytic reactions in residues are those that resulted from heating at very high temperatures, usually between 300° and 350°. Pyrolytic GC/MS (or py-GC/MS) is a version of GC/MS where the sample is heated to very high temperature before introduction to the GC in order to break down complicated components into smaller ones.

PUFAs: polyunsaturated fatty acids. Fatty acids containing more than one double bond. They oxidize more easily than unsaturated or monounsaturated fatty acids.

Quadrupole mass spectrometer: most common type of mass spectrometer for GC/MS and HPLC/MS applications. It is reasonably priced and permits good sensitivity across a range of mass weights. The name derives from four metallic rods with oscillating electric fields that permit the separation and detection of a range of mass weights. The oscillation in the electric fields means that quadrupole MSes usually use a scan mode to measure a wide range of mass weights, which take a short but measurable period of time, as described in chapter 4.

Quantification: determination of the quantities of a known compounds present in a residue using the linearity of a known compound as measured with an external calibration curve. The most accurate method of measuring the amount of compounds in a residue but is only practicable for targeted analysis. Quantification results are generally given in μg lipid/g pottery sherd.

Retention gap: short length (1–5 m, as capillary columns are generally 15 m or longer) of uncoated, deactivated capillary column that is often placed between an on-column injector and the capillary column. When too many unevaporated or undissolved compounds build up in the retention gap (see on-column injector), the retention gap can be replaced, prolonging the life of the capillary column.

Reverse phase: type of column in HPLC/MS that is most commonly used in residue analysis. Reverse phase columns have a nonpolar stationary phase, and polar mobile phases are used as solvents.

Saturated compounds: compounds with no double bonds.

Semiquantification: determination of the approximate quantities of compounds in a sample, including both knowns and unknowns, by using an internal standard. It assumes the linearity of all compounds in the residue is the same as that of the internal standard, which is untrue, because each compound will have slightly different linearity. Therefore, this process is less accurate than a true quantification using an external calibration curve but can be applied to all compounds in a residue at once, including unknown compounds. As a result, semiquantification is most commonly used in untargeted residue analysis, which includes a large number of compounds in each sample. Semiquantification results are generally given in μg lipid/g pottery sherd.

SIM: selected ion monitoring, a technique in which a quadrupole mass spectrometer is set to measure only a limited number of targeted molecular weights. This allows much more sensitive detection of the targeted weights, but no data is collected on the untargeted weights.

Solvent blank: in some labs, the solvent blank is the same as the procedural blank. In others, the solvent blank is solvent that is injected into an instrument to look for the possibility of ghost peaks.

SOPs: standard operating procedure (or protocol). These are the written instructions for laboratory procedure that every lab will have on file. They go through all laboratory processes in detail to allow consistent training of students or new hires, to prevent omission of important steps, and to allow a safety review of lab procedures. If any sampling or swabbing needs to take place away from the residue lab, you should ask for the relevant SOPs.

Split/splitless injector: type of injector in a GC system in which the sample is injected into an injector port, generally held at a high temperature (see Injector temperature). When injected, the sample can either all be introduced to the column to optimize detection of compounds (splitless injection) or some given fraction of the sample can be introduced to the column, while the rest is vented (split injection). Split injection is used for samples with large amounts of compounds that could overwhelm the column and/or detector.

TAGs: triacylglycerols, a molecule with three fatty acids attached to a glycerol backbone. These compounds make up the majority of lipids (fats and oils) from nearly all sources.

TAH fraction: total acidic hydrolysis fraction. This is a residue extracted from a sherd using the acidic methanol extraction. It contains all the lipids from a residue, but all hydrolyzed, which breaks acyl lipids and wax esters down into their component parts, as well as other complex molecules with an oxygen linkage.

TLE: total lipid extract. All the lipids removed from the archaeological matrix (pottery or carbon) during laboratory procedures.

TLE fraction: residue sample that includes all lipids from an extraction. This is usually used to describe the lipids extracted using chloroform/methanol extraction. It is also sometimes used to refer to a TAH fraction, which also includes all lipids from an extraction but in hydrolyzed form.

TMS ester: trimethylsilyl ester, a compound type often produced during derivatization of compounds with a hydroxyl group.

TMTD: 4,8,12 trimethyltridecanoic acid, an isoprenoid fatty acid often used as a marker for fish or shellfish presence in a residue.

TOF mass spectrometer: Time-of-flight mass spectrometer. This type of mass spectrometer accelerates mass fragments through an electric field and measures the time taken for each fragment to travel a known distance. This allows the determination of the weight of the fragment, as heavier fragments are slower than lighter fragments. The TOF MS scans all fragment weights simultaneously, as opposed to scanning a range of weights (as in the quadrupole), so TOF MS is more sensitive than quadrupole MS, essentially always producing SIM-like sensitivity while measuring the complete range of mass fragments. This type of mass spectrometer is noticeably more expensive than quadrupole, however, and is often saved for applications that require a high degree of sensitivity or with extraordinarily complex chromatograms.

TPP: triphenyl phosphate, a common plasticizer.

Trimethylsilyl esters: derivatized compound in which a trimethylsilyl group replaces a hydrogen in hydroxyl groups, so that –OH becomes $-OSi(CH_3)$. This makes the compounds more volatile and produces more sensitive, quantifiable analysis. It is the preferred derivatization for neutral compounds, such as alkanols and sterols, because these compounds do not easily form methyl esters during derivatization, and because it often allows a better understanding of a compound's structure. It adds three modern carbons to archaeological compounds when they undergo CSIA, which is why there are often larger uncertainties in CSIA results of neutral compounds than of fatty acids.

Trophic shift: slight change in stable isotope values in a compound as it travels up the food chain. Trophic shifts occur when a compound is modified, synthesized, or in some way changed within the body of the creature that ingested the compound. The biochemical reaction within the body tends to preferentially use one or another isotope of an element, resulting in a trophic shift. These shifts are particularly pronounced in stable nitrogen isotopes in bone collagen, because the collagen must be synthesized from dietary proteins. This is why bone collagen can often be used to determine approximately how high on the food chain an individual is eating. If a compound does not undergo any modification after ingestion, there will be no trophic shift; this often happens with common lipids.

Tuning: quadrupole mass spectrometers sometimes suffer from slight variation in electrical current and detector positioning, resulting in a small drift in the measurement of mass weights. Tuning corrects for this drift. Each lab will have a policy for tuning the MS, often related to EPA standards for analytical laboratories.

Turbo pump: mass spectrometers require high vacuum to function. The turbo pump is the fine pump inside the instrument responsible for keeping the MS at workable vacuum levels. Turbo pump failures are loud, expensive, and occasionally catastrophic.

UCM: Unresolved Complex Mixture, the large, unresolved hump produced by crude oil in a standard GC/MS analysis.

Undeveloped soil: soils composed primarily of sediment and without much organic material present. Degree of development of the soil affects the freshwater reservoir effect and may also affect residue preservation during deposition.

Volatilize: evaporate at a lower temperature. Derivatizing agents allow compounds to volatilize at temperatures that are easier for a GC injector and column to reach.

REFERENCES CITED

Ackman, R. G.

1964 Structural Homogeneity in Unsaturated Fatty Acids of Marine Lipids. A Review. *Journal of the Fisheries Research Board of Canada* 21(2):247–254.

Ackman, R. G., and S. N. Hooper

1970 Branched-Chain Fatty Acids of Four Fresh-Water Fish Oils. *Comparative Biochemistry and Physiology* 32(1):117–125. DOI:10.1016/0010–406X(70)90160-X.

Ackman, R. G., S. N. Hooper, and P. J. Ke

1971 The Distribution of Saturated and Isoprenoid Fatty Acids in the Lipids of Three Species of Molluscs, *Littorina littorea*, *Crassostrea virginica*, and *Venus mercenaria*. *Comparative Biochemistry and Physiology Part B: Comparative Biochemistry* 39(3):579–587.

Adamant, Ashley

2018 How to Eat a Pine Tree (and Other Conifers). *Practical Self-Reliance* (blog), December 11, 2018. https://practicalselfreliance.com.

Alt, Susan M.

2002 Identities, Traditions, and Diversity in Cahokia's Uplands. *Midcontinental Journal of Archaeology* 27(2):217–235.

Ambrose, Stanley H.

1991 Effects of Diet, Climate, and Physiology on Nitrogen Isotope Abundances in Terrestrial Foodwebs. *Journal of Archaeological Science* 18(3):293–317.

Anderson, Shelby L., Shannon Tushingham, and Tammy Y. Buonasera

2017 Aquatic Adaptations and the Adoption of Arctic Pottery Technology: Results of Residue Analysis. *American Antiquity* 82(3):452–479. DOI:10.1017/aaq.2017.8.

Aniszewski, Tadeusz

2007 *Alkaloids—Secrets of Life: Alkaloid Chemistry, Biological Significance, Applications and Ecological Role*. Elsevier, Oxford.

Assad-Bustillos, Melissa, Mariana Ramírez-Gilly, Alberto Tecante, and Leandro Chaires-Martínez

2014 Physicochemical, Functional, Thermal, and Rheological Characterization of Starch from Huauzontle Seeds (*Chenopodium berlandieri* spp. *nuttalliae*). *Agrociencia* 48(8):789–803.

Austin, Amy T., and P. M. Vitousek
1998 Nutrient Dynamics on a Precipitation Gradient in Hawai'i. *Oecologia* 113(4):519–529.

Avigan, Joel, and Max Blumer
1968 On the Origin of Pristane in Marine Organisms. *Journal of Lipid Research* 9(3):350–352.

Badler, Virginia R., Patrick E. McGovern, and Rudolph H. Michel
1990 Drink and Be Merry! Infrared Spectroscopy and Ancient Near Eastern Wine. In *Organic Contents of Ancient Vessels: Materials Analysis and Archaeological Investigation*, edited by W. Biers and P. McGovern, pp. 25–36. MASCA Research Papers in Science and Archaeology Vol. 7. University of Pennsylvania Press, Philadelphia.

Ballentine, Donna C., Stephen A. Macko, Vaughan C. Turekian, William P. Gilhooly, and Bice Martincigh
1996 Compound Specific Isotope Analysis of Fatty Acids and Polycyclic Aromatic Hydrocarbons in Aerosols: Implications for Biomass Burning. *Organic Geochemistry* 25(102):97–104.

Barnard, Hans, Alek N. Dooley, Gregory Areshian, Boris Gasparyan, and Kym F. Faull
2011 Chemical Evidence for Wine Production around 4000 BCE in the Late Chalcolithic Near Eastern Highlands. *Journal of Archaeological Science* 38(5):977–984.

Baumann, Timothy, Tammy Gerke, and Eleanora Reber
2013 Sun Circles and Science: Negative Painted Pottery from Angel Mounds (12Vg1). *Midcontinental Journal of Archaeology* 38(2):219–244.

Beaman, Thomas E., Jr., Vincent H. Melomo, J. Matthew Compton, Jennifer L. Gabriel, Matthew T. Kerr, Daniel J. Polito, Eleanora A. Reber, Alexandria D. Salisbury, Hannah P. Smith, and Linda F. Stine
2016 *"At Night We Sleep in a Shanty I Have Constructed of Planks, Logs, and Sand . . . On a Pile of Pine Straw": Archaeological Excavations of the Fort Anderson Overflow Barracks Area West of Battery A (31Bw376**12) and Metal Detector Survey of a Suspected Barracks Area West of Battery B (31Bw376**7) at Brunswick Town/Fort Anderson State Historic Site.* Wake Technical Community College, Raleigh, North Carolina.

Berlant, Anthony, and Evan M. Maurer
2017 *Decoding Mimbres Painting: Ancient Ceramics of the American Southwest.* Prestel, Los Angeles.

Bianchi, Giorgio, Pinarosa Avato, and Francesco Salamini
1984 Surface Waxes from Grain, Leaves, and Husks of Maize (*Zea mays* L.). *Cereal Chemistry* 61(1):45–47.

Blanco-Zubiaguirre, Laura, Maitane Olivares, Kepa Castro, Jose Antonio Carrero, Carlos García-Benito, José Ángel García-Serrano, Julián Pérez-Pérez, and Josefina Pérez-Arantegui
2019 Wine Markers in Archeological Potteries: Detection by GC-MS at Ultratrace Levels. *Analytical and Bioanalytical Chemistry* 411:6711–6722.

Bragagnolo, N., and D. B. Rodriguez-Amaya
2001 Total Lipid, Cholesterol, and Fatty Acids of Farmed Freshwater Prawn (*Macrobrachium rosenbergii*) and Wild Marine Shrimp (*Penaeus brasiliensis, Penaeus*

schimitti, Xiphopenaeus kroyeri). *Journal of Food Composition and Analysis* 14(4):359–369.

Brassell, S. C., G. Eglinton, and J. R. Maxwell

1983 The Geochemistry of Terpenoids and Steroids. *Biochemical Society Transactions* 11(5):575–586. DOI:10.1042/bst0110575

Brown, Louise D., and Carl Heron

2005 Presence or Absence: A Preliminary Study into the Detection of Fish Oils in Ceramics. In *The Zooarchaeology of Milk and Fats*, edited by J. Milville and A. Outram, pp. 67–76. Oxbow Books, Oxford.

Buonasera, Tammy

2007 Investigating the Presence of Ancient Absorbed Organic Residues in Groundstone Using GC–MS and Other Analytical Techniques: A Residue Study of Several Prehistoric Milling Tools from Central California. *Journal of Archaeological Science* 34(9):1379–1390.

Carmody, Stephen B., Megan C. Kassabaum, Ryan K. Hunt, Natalie Prodanovich, Hope Elliott, and Jon Russ

2018 Residue Analysis of Smoking Pipe Fragments from the Feltus Archaeological Site, Southeastern North America. *Journal of Archaeological Science: Reports* 17:640–649.

Cernusak, Lucas A., Nerea Ubierna, Klaus Winter, Joseph A. M. Holtum, John D. Marshall, and Graham D. Farquhar

2013 Environmental and Physiological Determinants of Carbon Isotope Discrimination in Terrestrial Plants. *New Phytologist* 200:950–965. DOI:10.1111/nph.12423

Charters, S., R. P. Evershed, L. J. Goad, C. Heron, and P. Blinkhorn

1993b Identification of an Adhesive Used to Repair a Roman Jar. *Archaeometry* 35(1):91–101.

Charters, S., R. P. Evershed, L. J. Goad, A. Leyden, P. W. Blinkhorn, and V. Denham

1993a Quantification and Distribution of Lipid in Archaeological Ceramics: Implications for Sampling Potsherds for Organic Residue Analysis and the Classification of Vessel Use. *Archaeometry* 35(2):211–223.

Chibnall, Albert Charles, Stephen Harvey Piper, Alfred Pollard, Ernest Frank Williams, and Prem Nath Sahai

1934 The Constitution of the Primary Alcohols, Fatty Acids, and Paraffins Present in Plant and Insect Waxes. *Biochemical Journal* 28(6):2189–2208.

Chisholm B. S., D. E. Nelson, and H. P. Schwarcz

1982 Stable-Carbon Isotope Ratios as a Measure of Marine versus Terrestrial Protein in Ancient Diets. *Science* 216(4550):1131–1132. DOI:10.1126/science.216.4550.1131.

Choi, Woo-Jung, Hee-Myong Ro, and Erik A. Hobbie

2003 Patterns of Natural ^{15}N in Soils and Plants from Chemically and Organically Fertilized Uplands. *Soil Biology and Biochemistry* 35(11):1493–1500. DOI:10.1016/S0038-0717(03)00246-3.

Choy, Kyungcheol, Ben A. Potter, Holly J. McKinney, Joshua D. Reuther, Shiway W. Wang, and Mathew J. Wooller

2016 Chemical Profiling of Ancient Hearths Reveals Recurrent Salmon Use in Ice Age Beringia. *Proceedings of the National Academy of Sciences* 113(35):9757–9762.

Condamin, J., F. Formenti, M. O. Metais, M. Michel, and P. Blond
1976 The Application of Gas Chromatography to the Tracing of Oil in Ancient Amphorae. *Archaeometry* 18(2):195–201.

Copley, Mark S., Rob Berstan, Stephanie N. Dudd, Gordon Docherty, Anna J. Mukherjee, V. Straker, S. Payne, and Richard P. Evershed
2003 Direct Chemical Evidence for Widespread Dairying in Prehistoric Britain. *Proceedings of the National Academy of Sciences USA* 100(4):1524–1529.

Copley, Mark S., Fabricio A. Hansel, Karim Sadr, and Richard P. Evershed
2004 Organic Residue Evidence for the Processing of Marine Animal Products in Pottery Vessels from the Pre-colonial Archaeological Site of Kasteelberg D East, South Africa. *South African Journal of Science* 100:279–283.

Copley, M. S., R. Berstan, S. N. Dudd, V. Straker, S. Payne, and R. P. Evershed
2005 Dairying in Antiquity. I. Evidence from Absorbed Lipid Residues Dating to the British Iron Age. *Journal of Archaeological Science* 32(4):485–503.

Corr, Lorna T., Michael P. Richards, Susan Jim, Stanley H. Ambrose, Alexander Mackie, Owen Beattie, and Richard P. Evershed
2008 Probing Dietary Change of the Kwädąy Dän Ts'ìnchį Individual, an Ancient Glacier Body from British Columbia: I. Complementary Use of Marine Lipid Biomarker and Carbon Isotope Signatures as Novel Indicators of a Marine Diet. *Journal of Archaeological Science* 35(8):2102–2110.

Correa-Ascencio, Marisol, and Richard P. Evershed
2014 High Throughput Screening of Organic Residues in Archaeological Potsherds Using Direct Acidified Methanol Extraction. *Analytical Methods* 6(5):1330–1340.

Correa-Ascencio, Marisol, Ian G. Robertson, Oralia Cabrera-Cortés, Rubén Cabrera-Castro, and Richard P. Evershed
2014 Pulque Production in Prehispanic Mesoamerica. *Proceedings of the National Academy of Sciences* 111(39):14223–14228. DOI:10.1073/pnas.1408339111.

Courel, Blandine, Philippe Schaeffer, Pierre Adam, Estelle Motsch, Quentin Ebert, Emile Moser, Clément Féliu, Stefano M. Bernasconi, Irka Hajdas, Damien Ertlen, and Dominique Schwartz
2017 Molecular, Isotopic, and Radiocarbon Evidence for Broomcorn Millet Cropping in Northeast France since the Bronze Age. *Organic Geochemistry* 110:13–24.

Cowgill, George L.
2015 Some Things I Hope You Will Find Useful Even If Statistics Isn't Your Thing. *Annual Reviews of Anthropology* 44:1–14.

Craig, Oliver E., Richard B. Allen, Anu Thompson, Rhiannon E. Stevens, Valerie J. Steele, and Carl Heron
2012 Distinguishing Wild Ruminant Lipids by Gas Chromatography/Combustion/Isotope Ratio Mass Spectrometry. *Rapid Communications in Mass Spectrometry* 26:2359–2364.

Craig, Oliver E., M. Forster, Søren H. Andersen, E. Koch, P. Crombé, N. J. Milner, Ben Stern, G. N. Bailey, and Carl P. Heron
2007 Molecular and Isotopic Demonstration of the Processing of Aquatic Products in Northern European Prehistoric Pottery. *Archaeometry* 49(1):135–152.

Craig, Oliver E., Gordon D. Love, Sven Isaksson, Gillian Taylor, and Colin E. Snape

2004 Stable Carbon Isotopic Characterisation of Free and Bound Lipid Constituents of Archaeological Ceramic Vessels Released by Solvent Extraction, Alkaline Hydrolysis, and Catalytic Hydropyrolysis. *Journal of Analytical and Applied Pyrolysis* 71(2):613–634. DOI:10.1016/j.jaap.2003.09.001.

Craig, O. E., H. Saul, A. Lucquin, Y. Nishida, K. Taché, L. Clarke, A. Thompson, D. T. Altoft, J. Uchiyama, M. Ajimoto, K. Gibbs, S. Isaksson, C. P. Heron, and P. Jordan

2013 Earliest Evidence for the Use of Pottery. *Nature* 496(7445): 351–354.

Craig, Oliver E., Lisa-Marie Shillito, Umberto Albarella, Sarah Viner-Daniels, Ben Chan, Ros Cleal, Robert Ixer, Mandy Jay, Pete Marshall, Ellen Simmons, Elizabeth Wright, and Mike Parker Pearson

2015 Feeding Stonehenge: Cuisine and Consumption at the Late Neolithic Site of Durrington Walls. *Antiquity* 89(347):1096–1109.

Craig, Oliver E., Val J. Steele, Anders Fischer, Sonke Hartz, Søren H. Andersen, Paul Donohoe, Aikaterini Glykou, Hayley Saul, D. Martin Jones, Eva Koch, and Carl P. Heron

2011 Ancient Lipids Reveal Continuity in Culinary Practices across the Transition to Agriculture in Northern Europe. *Proceedings of the National Academy of Sciences* 108(440):17910–17915.

Cramp, Lucy J. E., Jennifer Jones, Alison Sheridan, Jessica Smyth, Helen Whelton, Jacqui Mulville, Niall Sharples, and Richard P. Evershed

2014 Immediate Replacement of Fishing with Dairying by the Earliest Farmers of the Northeast Atlantic Archipelagos. *Proceedings of the Royal Society B* 281(1780):20132372. DOI:10.1098/rspb.2013.2372.

Crown, Patricia L., Thomas E. Emerson, Jiyan Gu, W. Jeffrey Hurst, Timothy R. Pauketat, and Timothy Ward

2012 Ritual Black Drink Consumption at Cahokia. *Proceedings of the National Academy of Sciences* 109(35):13944–13949.

de Paula Lima, Juliana, and Ariana Farah

2019 Methylxanthines in Stimulant Foods and Beverages Commonly Consumed in Brazil. *Journal of Food Composition and Analysis* 78:75–85. DOI:10.1016/j.jfca.2019.02.001.

DeNiro, Michael J., and Samuel Epstein

1978 Influence of Diet on the Distribution of Carbon Isotopes in Animals. *Geochimica et Cosmochimica Acta* 42:495–506.

Downing, Donald T.

1976 Mammalian Waxes. In *Chemistry and Biochemistry of Natural Waxes*, edited by P. E. Kolattukudy, pp. 18–48. Elsevier, Amsterdam.

Drieu, L., M. Horgnies, D. Binder, P. Pétrequin, A.-M. Pétrequin, K. Peche-Quilichini., T. Lachenal, and M. Regert

2019 Influence of Porosity on Lipid Preservation in the Wall of Archaeological Pottery. *Archaeometry* 61:1081–1096. doi.org/10.1111/arcm.12479.

Drieu, Léa, Maxime Rageot, Nathan Wales, Ben Stern, Jasmine Lundy, Maximilian Zerrer, Isabella Gaffney, Manon Bondetti, Cynthianne Spiteri, Jane Thomas-Oates, and Oliver E. Craig

2020 Is It Possible to Identify Ancient Wine Production Using Biomolecular

Approaches? *STAR: Science and Technology of Archaeological Research* 6(1):16–29. DOI:10.1080/20548923.2020.1738728.

Dudd, Stephanie N., and Richard P. Evershed

1998 Direct Demonstration of Milk as an Element of Archaeological Economies. *Science* 282(5393):1478–1481. DOI:10.1126/science.282.5393.1478.

Dudd, Stephanie N., Richard P. Evershed, and Alex M. Gibson

1999 Evidence for Varying Patterns of Exploitation of Animal Products in Different Prehistoric Pottery Traditions Based on Lipids Preserved in Surface and Absorbed Residues. *Journal of Archaeological Science* 26(12):1473–1482.

Durzan, Don J.

2009 Arginine, Scurvy, and Cartier's "Tree of Life." *Journal of Ethnobiology and Ethnomedicine* 5:5. DOI:10.1186/1746-4269-5-5.

Echeverría, Javier, María Teresa Planella, and Hermann M. Niemeyer

2014 Nicotine in Residues of Smoking Pipes and Other Artifacts of the Smoking Complex from an Early Ceramic Period Archaeological Site in Central Chile. *Journal of Archaeological Science* 44:55–60.

Edwards, Adam L., and Bradley C. Bennett

2005 Diversity of Methylxanthine Content in *Ilex cassine* L. and *Ilex vomitoria* Ait.: Assessing Sources of the North American Stimulant Cassina. *Economic Botany* 59(3):275–285.

Eerkens, Jelmer

2002 The Preservation and Identification of Piñon Resins by GC-MS in Pottery from the Western Great Basin. *Archaeometry* 44(1):95–105.

2005 GC-MS Analysis and Fatty Acid Ratios of Archaeological Potsherds from the Western Great Basin of North America. *Archaeometry* 47(1):83–102.

Emerson, Thomas E., Kristin M. Hedman, and Mary L. Simon

2005 Marginal Horticulturalists or Maize Agriculturalists? Archaeobotanical, Paleopathological, and Isotopic Evidence Relating to Langford Tradition Maize Consumption. *Midcontinental Journal of Archaeology* 30(1):67–118.

Evershed, Richard P.

1993 Biomolecular Archaeology and Lipids. *World Archaeology* 25(1):74–93.

2008a Experimental Approaches to the Interpretation of Absorbed Organic Residues in Archaeological Ceramics. *World Archaeology* 40(1):26–47.

2008b Organic Residue Analysis in Archaeology: the Archaeological Biomarker Revolution. *Archaeometry* 50(6):895–924.

Evershed, Richard P., and Stephanie Charters

1995 Simulating the Degradation of Animal Fats in Archaeological Ceramics. *Organic Geochemistry: Developments and Applications to Energy, Climate, Environment, and Human History*:697–698. A.I.G.O.A, San Sebastián, Spain.

Evershed, R. P., M. S. Copley, L. Dickson, and F. A. Hansel

2008b Experimental Evidence for the Processing of Marine Animal Products and Other Commodities Containing Polyunsaturated Fatty Acids in Pottery Vessels. *Archaeometry* 50(1):101–113.

Evershed, Richard P., Carl Heron, and L. John Goad

1990 Analysis of Organic Residues of Archaeological Origin by High-Temperature

Gas Chromatography and Gas Chromatography–Mass Spectrometry. *Analyst* 115:1339–1342.

1991 Epicuticular Wax Components Preserved in Potsherds as Chemical Indicators of Leafy Vegetables in Ancient Diets. *Antiquity* 65(248):540–544. DOI:10.1017/S0003598X00080145.

Evershed, Richard P., Stephanie N. Dudd, Mark S. Copley, and Anna Mukherjee

2002 Identification of Animal Fats via Compound Specific $\delta^{13}C$ Values of Individual Fatty Acids: Assessments of Results for Reference Fats and Lipid Extracts of Archaeological Pottery Vessels. *Documenta Praehistorica* 29:73–96. DOI:10.4312/dp.29.7.

Evershed, R. P., H. R. Mottram, S. N. Dudd, S. Charters, A. W. Stott, G. J. Lawrence, A. M. Gibson, A. Conner, P. Blinkhorn, and V. Reeeves

1997 New Criteria for the Identification of Animal Fats Preserved in Archaeological Pottery. *Naturwissenschaften* 84(9):402–406.

Evershed, Richard P., Sebastian Payne, Andrew G. Sherratt, Mark S. Copley, Jennifer Coolidge, Duska Urem-Kotsu, Kostas Kotsakis, Mehmet Özdoğan, Aslý E. Özdoğan, Olivier Nieuwenhuyse, Peter M. M. G. Akkermans, Douglass Bailey, Radian-Romus Andeescu, Stuart Campbell, Shahina Farid, Ian Hodder, Nurcan Yalman, Mihriban Özbaşaran, Erhan Bıçakcı, Yossef Garfinkel, Thomas Levy, and Margie M. Burton

2008a Earliest Date for Milk Use in the Near East and Southeastern Europe Linked to Cattle Herding. *Nature* 455(7212):528–531. DOI:10.1038/nature07180.

Evershed, R. P., A. W. Stott, A. Raven, S. N. Dudd, S. Charters, and A. Leyden

1995 Formation of Long-Chain Ketones in Ancient-Pottery Vessels by Pyrolysis of Acyl Lipids. *Tetrahedron Letters* 36(48):8875–8878.

Farrell, Thomas F. G., Peter Jordan, Karine Taché, Alexandre Lucquin, Kevin Gibbs, Ana Jorge, Kate Britton, Oliver E. Craig, and Rick Knecht

2014 Specialized Processing of Aquatic Resources in Prehistoric Alaskan Pottery?: A Lipid-Residue Analysis of Ceramic Sherds from the Thule-Period Site of Nunalleq, Alaska. *Arctic Anthropology* 51(1):86–100.

Fernandes, Ricardo, Yvette Eley, Marek Brabec, Alexandre Lucquin, Andrew Millard, and Oliver E. Craig

2018 Reconstruction of Prehistoric Pottery Use from Fatty Acid Carbon Isotope Signatures Using Bayesian Inference. *Organic Geochemistry* 117:31–42.

Fernandes, Ricardo, Andrew R. Millard, Marek Brabec, Marie-Josée Nadeau, and Pieter Grootes

2014 Food Reconstruction Using Isotopic Transferred Signals (FRUITS): A Bayesian Model for Diet Reconstruction. *PLoS ONE* 9(2):e87436.

Fewkes, Vladimir J.

1944 Catawba Pottery-Making, with Notes on Pamunkey Pottery-Making, Cherokee Pottery-Making, and Coiling. *Proceedings of the American Philosophical Society* 88(2):69–124.

Fischer, Anders, and Jan Heinemeier

2003 Freshwater Reservoir Effect in ^{14}C Dates of Food Residue on Pottery. *Radiocarbon* 45(3):449–466.

Frick, A. A., G. Chidlow, J. V. Goodpaster, S. W. Lewis, and W. van Bronswijk

2016 Monitoring Compositional Changes of the Lipid Fraction of Fingermark Residues Deposited on Paper during Storage. *Forensic Chemistry* 2:29–36. DOI:10.1016/j.forc.2016.09.001.

Fritz, Gayle J.

2019 *Feeding Cahokia: Early Agriculture in the North American Heartland*. University of Alabama Press, Tuscaloosa.

Fritz, Gayle J., and T. R. Kidder

1993 Recent Investigations into Prehistoric Agriculture in the Lower Mississippi Valley. *Southeastern Archaeology* 12:1–14.

Fritz, Gayle J., Virginia Drywater Whitekiller, and James W. McIntosh

2001 Ethnobotany of Ku-nu-che: Cherokee Hickory Nut Soup. *Journal of Ethnobiology* 21(2):1–27.

Ganzarolli, Giovanna, Michelle Alexander, Alexandra Chavarria Arnau, and Oliver E. Craig

2018 Direct Evidence from Lipid Residue Analysis for the Routine Consumption of Millet in Early Medieval Italy. *Journal of Archaeological Science* 96:124–130.

García, Pablo Anselmo, Alaíde Braga de Oliveira, and Ronan Batista

2007 Occurrence, Biological Activities, and Synthesis of Kaurane Diterpenes and Their Glycosides. *Molecules* 12(3):455–483. DOI:10.3390/12030455.

Gilmore, Melvin R.

1977 *Uses of Plants by the Indians of the Missouri River Region*. University of Nebraska Press, Lincoln.

Goldenberg, Larisa, Ronny Neumann, and Steve Weiner

2014 Microscale Distribution and Concentration of Preserved Organic Molecules with Carbon-Carbon Double Bonds in Archaeological Ceramics: Relevance to the Field of Residue Analysis. *Journal of Archaeological Science* 42:509–518.

Gremillion, Kristen J.

2004 Seed Processing and the Origins of Food Production in Eastern North America. *American Antiquity* 69(2):215–233. DOI:10.2307/4128417.

Guasch-Jané, Maria Rosa, Cristina Andrés-Lacueva, Olga Jáuregui, and Rosa M. Lamuela-Raventós

2006 First Evidence of White Wine in Ancient Egypt from Tutankhamun's Tomb. *Journal of Archaeological Science* 33(8):1075–1080. DOI:10.1016/j.jas.2005.11.012.

Gunstone, Frank D., John L. Harwood, and Fred B. Padley (editors)

1986 *The Lipid Handbook*. Chapman and Hall, London.

Guoping, Sun

2013 Recent Research on the Hemudu Culture and the Tianluoshan Site. In *A Companion to Chinese Archaeology*, edited by A. P. Underhill, pp. 555–573. Wiley-Blackwell, New York. DOI:10.1002/9781118325698.ch27.

Hammann, Simon, David J. Scurr, Morgan R. Alexander, and Lucy J. E. Cramp

2020 Mechanisms of Lipid Preservation in Archaeological Clay Ceramics Revealed by Mass Spectrometry Imaging. *Proceedings of the National Academy of Sciences* 117(26):14688–14693. DOI:10.1073/pnas.1922445117.

Hansel, Fabricio A., Mark S. Copley, Luiz A. S. Madureira, and Richard P. Evershed

2004 Thermally Produced ω-(o-Alkylphenyl)alkanoic Acids Provide Evidence for the

Processing of Marine Products in Archaeological Pottery Vessels. *Tetrahedron Letters* 45(14):2999–3002.

Harborne, Jeffrey B., and Francisco A. Tomas-Barberan
1991 *Ecological Chemistry and Biochemistry of Plant Terpenoids.* Proceedings of the Phytochemical Society of Europe Vol. 31. Clarendon, Oxford.

Hare, Vincent J., Emma Loftus, Amy Jeffrey, and Christopher Bronk Ramsey
2018 Atmospheric CO_2 Effect on Stable Carbon Isotope Composition of Terrestrial Fossil Archives. *Nature Communications* 9:252. DOI:10.1038/s41467-017-02691-x.

Hart, John P., William A. Lovis, Janet K. Schulenberg, and Gerald R. Urquhart
2007 Paleodietary Implications from Stable Carbon Isotope Analysis of Experimental Cooking Residues. *Journal of Archaeological Science* 34(5):804–813.

Hart, John, William Lovis, Gerald Urquhart, and Eleanora Reber
2013 Modeling Freshwater Reservoir Offsets on Radiocarbon-Dated Charred Cooking Residues. *American Antiquity* 78(3):536–552.

Hart, John P., Eleanora A. Reber, Robert G. Thompson, and Robert Lusteck
2008 Taking Variation Seriously: Testing the Steatite Mast-Processing Hypothesis with Microbotanical Data from the Hunter's Home Site, New York. *American Antiquity* 73(4):729–741. DOI:10.1017/S0002731600047375.

Harwood, John L.
2019 Algae: Critical Sources of Very Long-Chain Polyunsaturated Fatty Acids. *Biomolecules* 9(11):708. DOI:10.3390/biom9110708.

Heron, Carl, and Richard P. Evershed
1993 The Analysis of Organic Residues and the Study of Pottery Use. *Archaeological Method and Theory* 5:247–284.

Heron, Carl, Richard P. Evershed, and L. John Goad
1991 Effects of Migration of Soil Lipids on Organic Residues Associated with Buried Potsherds. *Journal of Archaeological Science* 18(6):641–659.

Heron, Carl, Richard P. Evershed, L. John Goad, and V. Denham
1989 New Approaches to the Analysis of Organic Residues from Archaeological Remains. In *Archaeological Sciences 1989: Proceedings of a Conference on the Application of Scientific Techniques to Archaeology*, edited by P. Budd, B. Chapman, C. Jackson, R. Janaway, and B. Ottaway, pp. 332–339. Oxbow, Oxford.

Heron, Carl, Shinya Shoda, Adrià Breu Barcons, Janusz Czebreszuk, Yvette Eley, Marise Gorton, Wiebke Kirleis, Jutta Kneisel, Alexandre Lucquin, Johannes Müller, Yastami Nishida, Joon-ho Son, and Oliver E. Craig
2016 First Molecular and Isotopic Evidence of Millet Processing in Prehistoric Pottery Vessels. *Scientific Reports* 6(1):38767.

Huang Zih-Rou, Yin-Ku Lin, and Jia-You Fang
2009 Biological and Pharmacological Activities of Squalene and Related Compounds: Potential Uses in Cosmetic Dermatology. *Molecules* 14(1):540–554. DOI:10.3390/molecules14010540.

Hudson, Charles M. (editor)
2004 *Black Drink: A Native American Tea.* University of Georgia Press, Athens.

Hurst, W. J., R. Martin Jr., S. Tarka Jr., and G. C. Hall

1989 Authentication of Cocoa in Maya Vessels Using High-Performance Liquid Chromatographic Techniques. *Journal of Chromatography* 466:279–289.

Isaksson, Sven, and Fredrik Hallgren

2012 Lipid Residue Analyses of Early Neolithic Funnel-Beaker Pottery from Skogsmossen, Eastern Central Sweden, and the Earliest Evidence of Dairying in Sweden. *Journal of Archaeological Science* 39(12):3600–3609.

Jacob, Jeremy, Jean-Robert Disnar, Mohammed Boussafir, Ana Luiza Spadano Albuquerque, Abdelfettah Sifeddine, and Bruno Turcq

2005 Pentacyclic Triterpene Methyl Ethers in Recent Lacustrine Sediments (Lagoa do Caco, Brazil). *Organic Geochemistry* 36(3):449–461.

Joye, Samantha B.

2015 Deepwater Horizon, 5 Years On. *Science* 349(6248):592–593.

Keaveney, Evelyn M., and Paula J. Reimer

2012 Understanding the Variability in Freshwater Radiocarbon Reservoir Offsets: A Cautionary Tale. *Journal of Archaeological Science* 39(5):1306–1316.

Kidder T. R.

2002 Woodland Period Archaeology of the Lower Mississippi Valley. In *The Woodland Southeast*, edited by David G. Anderson and Robert C. Mainfort Jr, pp. 66–90. University of Alabama Press, Tuscaloosa.

King, Adam, Terry Powis, Kong Cheong, Bobi Deere, Robert Pickering, Eric Singleton, and Nilesh Gaikwad

2018 Absorbed Residue Evidence for Prehistoric Datura Use in the American Southeast and Western Mexico. *Advances in Archaeological Practice* 6(4):312–327.

King, Adam, Terry G. Powis, Kong F. Cheong, and Nilesh W. Gaikwad

2017 Cautionary Tales on the Identification of Caffeinated Beverages in North America. *Journal of Archaeological Science* 85:30–40.

Kishore, Ravuri S. K., Sylvia Kiese, Stefan Fischer, Astrid Pappenberger, Ulla Grauschopf, and Hanns-Christian Mahler

2011 The Degradation of Polysorbates 20 and 80 and Its Potential Impact on the Stability of Biotherapeutics. *Pharmaceutical Research* 28:1194–1210.

Kohler, P., H. Fischer, and J. Schmitt

2010 Atmospheric $\delta^{13}CO_2$ and Its Relation to $p\ CO_2$ and Deep Ocean $\delta^{13}C$ during the Late Pleistocene. *Paleooceanography* 25(1):PA1213. DOI:10.1029/2008PA001703.

Kujawinski, Elizabeth B., Melissa C. Kido Soule, David L. Valentine, Angela K. Boysen, Krista Longnecker, and Molly C. Redmond

2011 Fate of Dispersants Associated with the Deepwater Horizon Oil Spill. *Environmental Science Technology* 45(4):1298–1306.

Langenheim, Jean H.

2003 *Plant Resins: Chemistry, Evolution, Ecology, and Ethnobotany*. Timber Press, Portland, Oregon.

Lazzara, Raimondo, Denise Fernandes, Melissa Faria, Jordi F. López, Romà Tauler, and Cinta Porte

2012 Changes in Lipid Content and Fatty Acid Composition along the Reproductive Cycle of the Freshwater Mussel *Dreissena polymorpha*: Its Modulation by Clofibrate Exposure. *Science of the Total Environment* 432:195–201.

Li, Jie, Li Pan, Joshua N. Fletcher, Wei Lv, Ye Deng, Michael A. Vincent, Jay P. Slack, T. Scott McCluskey, Zhonghua Jia, Mark Cushman, and A. Douglas Kinghorn

2014 In Vitro Evaluation of Potential Bitterness-Masking Terpenoids from the Canada Goldenrod (*Solidago canadensis*). *Journal of Natural Products* 77:1739–1743.

Lingbeck, Jody M., Corliss A. O'Bryan, Elizabeth M. Martin, Joshua P. Adams, and Philip G. Crandall

2015 Sweetgum: An Ancient Source of Beneficial Compounds with Modern Benefits. *Pharmacognosy Reviews* 9(17):1–11. DOI:10.4103/0973-7847.156307.

Listi, Ginesse A.

2011 Bioarchaeological Analysis of Diet during the Coles Creek Period in the Southern Lower Mississippi Valley. *American Journal of Physical Anthropology* 144:30–40. DOI:10.1002/ajpa.21364.

Longacre, William

1981 Kalinga Pottery: An Ethnoarchaeological Study. In *Pattern of the Past: Studies in Honour of David Clarke*, edited by I. Hodder, G. Isaac, and N. Hammond, pp. 49–66. Cambridge University Press, Cambridge.

Longacre, William A., and Taylor R. Hermes

2015 Rice Farming and Pottery Production among the Kalinga: New Ethnoarchaeological Data from the Philippines. *Journal of Anthropological Archaeology* 38:35–45.

Lucquin, Alexandre, André C. Colonese, Thomas F. G. Farrell, and Oliver E. Craig

2016 Utilising Phytanic Acid Diastereomers for the Characterisation of Archaeological Lipid Residues in Pottery Samples. *Tetrahedron Letters* 57(6):703–707.

Malainey, Mary E., Roman Przbylski, and Barbara L. Sherriff

1999a The Fatty Acid Composition of Native Food Plants and Animals of Western Canada. *Journal of Archaeological Science* 26(1):83–94.

1999b The Effects of Thermal and Oxidative Degradation on the Fatty Acid Composition of Food Plants and Animals of Western Canada: Implications for the Identification of Archaeological Vessel Residues. *Journal of Archaeological Science* 26(1):95–103.

Matikainen, Jorma, Seppo Kaltia, Maija Ala-Peijari, Ninna Petit-Gras, Kirsi Harju, Jaakko Heikkilä, Raija Yksjärvi, and Tapio Hase

2003 A Study of 1,5-Hydrogen Shift and Cyclization Reactions of an Alkali Isomerized Methyl Linolenoate. *Tetrahedron* 59(4):567–573. DOI:10.1016/S0040-4020(02)01513-2.

Maxwell, J. R., R. E Cox, G. Eglinton, C. T. Pillinger, R. G. Ackman, and S. N. Hooper

1973 Stereochemical Studies of Acyclic Isoprenoid Compounds—II The Role of Chlorophyll in the Derivation of Isoprenoid-Type Acids in a Lacustrine Sediment. *Geochimica et Cosmochimica Acta* 37(2):297–313. DOI:10.1016/0016-7037(73)90136-1.

Mazow, Laura, Susanne Grieve, and Anthony Kennedy

2014 Contamination in Organic Residue Analysis: A Cautionary Tale. *Journal of Eastern Mediterranean Archaeology* 2(2):90–109.

McElrath, Dale L., and Fred A. Finney

1987 *The George Reeves Site (11S650)*. FAI-270 Site Reports Vol. 15. University of Illinois Press, Urbana.

McGovern, Patrick E., Donald L. Glusker, Lawrence J. Exner, and Mary M. Voigt

1996 Neolithic Resinated Wine. *Nature* 381 (6582):480–481.

McGovern, Patrick E., Donald L. Glusker, Robert A. Moreau, Alberto Nuñez, Curt W. Beck, Elizabeth Simpson, Eric D. Butrym, Lawrence J. Exner, and Edith C. Stout

1999 A Funerary Feast Fit for King Midas. *Nature* 402:863–864. DOI:10.1038/47217.

McGovern, Patrick, Mindia Jalabadze, Stephen Batiuk, Michael P. Callahan, Karen E. Smith, Gretchen R. Hall, Eliso Kvavadze, David Maghradze, Nana Rusishvili, Laurent Bouby, Osvaldo Failla, Gabriele Cola, Luigi Mariani, Elisabetta Boaretto, Roberto Bacilieri, Patrice This, Nathan Wales, and David Lordkipanidze

2017 Early Neolithic Wine of Georgia in the South Caucasus. *Proceedings of the National Academy of Sciences of the United States of America* 114(48):E10309–E10318. DOI:10.1073/pnas.1714728114.

McLafferty, Fred W., and František Tureček

1993 *Interpretation of Mass Spectra*. 4th ed. University Science Books, Mill Valley, California.

Meyer, Buffy M., Puspa L. Adhikari, Gregory M. Olson, Edward B. Overton, and Martin Scott Miles

2018 Louisiana Coastal Marsh Environments and MC252 Oil Biomarker Chemistry. In *Oil Spill Environmental Forensics Case Studies*, edited by Scott A. Stout and Zhendi Wang, pp. 737–756. Butterworth-Heinemann, Oxford.

Michael, William R.

1966 Thermal Reactions of Methyl Linoleate. II. The Structure of Aromatic C_{18} Methyl Esters. *Lipids* 1:359–364. DOI:10.1007/BF02532681.

Michael, William R., J. Craig Alexander, and Neil R. Artman

1966 Thermal Reactions of Methyl Linoleate. I. Heating Conditions, Isolation Techniques, Biological Studies, and Chemical Changes. *Lipids* 1:353–358. DOI:10.1007/BF02532680.

Michel, Rudolph H., Patrick E. McGovern, and Virginia R. Badler

1993 The First Wine and Beer: Chemical Detection of Ancient Fermented Beverages. *Analytical Chemistry* 65(8):408–413.

Milanich, Jerald T.

2004 Origins and Prehistoric Distributions of the Black Drink and the Ceremonial Shell Drinking Cup. In *Black Drink: A Native American Tea*, edited by Charles M. Hudson, pp. 83–119. University of Georgia Press, Athens.

Mills, John S., and Raymond White

1977 Natural Resins of Art and Archaeology: Their Sources, Chemistry, and Identification. *Studies in Conservation* 22(1):12–31.

Moerman, Daniel E.

1998 *Native American Ethnobotany*. Timber Press, Portland, Oregon.

Mohlenbrock, Robert H.

1986 *Guide to the Vascular Flora of Illinois*. Southern Illinois University Press, Carbondale.

Mukherjee, Anna J., Robert Berstan, Mark S. Copley, Alex M. Gibson, and Richard P. Evershed

2007 Compound-Specific Stable Carbon Isotopic Detection of Pig Product Processing in British Late Neolithic Pottery. *Antiquity* 81(313):743–754.

Namera, Akira, Mikio Yashiki, Yasuo Hirose, Seiichi Yamaji, Tadato Tani, and Tohru Kojima
2002 Quantitative Analysis of Tropane Alkaloids in Biological Materials by Gas Chromatography–Mass Spectrometry. *Forensic Science International* 130(1):34–43.
National Institute of Standards
2014 NIST Standard Reference Database 1A v17. https://www.nist.gov.
Newton, Teresa J., Caryn C. Vaughn, Daniel E. Spooner, S. Jerrine Nichols, and Michael T. Arts
2013 Profiles of Biochemical Tracers in Unionid Mussels across a Broad Geographical Range. *Journal of Shellfish Research* 32(2):497–507.
Nuss, Emily T., and Sherry A. Tanumihardjo
2010 Maize: A Paramount Staple Crop in the Context of Global Nutrition. *Comprehensive Reviews in Food Science and Food Safety* 9:417–436.
Olsson, Monika, and Sven Isaksson
2008 Molecular and Isotopic Traces of Cooking and Consumption of Fish at an Early Medieval Manor site in Eastern Middle Sweden. *Journal of Archaeological Science* 35(3):773–780.
Onami, Tetsuo, and Hitoshi Kanazawa
1996 A Simple Method for Isolation of Caffeine from Black Tea Leaves: Use of a Dichloromethane-Alkaline Water Mixture as an Extractant. *Journal of Chemical Education* 73(6):556.
Pauketat, Timothy
2009 *Cahokia: Ancient America's Great City on the Mississippi*. Penguin Books, New York.
Perruchini, E., C. Glatz, M. M. Hald, J. Casana, and J. L. Toney
2018 Revealing Invisible Brews: A New Approach to the Chemical Identification of Ancient Beer. *Journal of Archaeological Science* 100:176–190.
Philippsen, Bente
2013 The Freshwater Reservoir Effect in Radiocarbon Dating. *Heritage Science* 1(1):24.
Polissar, Pratigya J., and William J. D'Andrea
2014 Uncertainty in Paleohydrologic Reconstructions from Molecular δD Values. *Geochimica et Cosmochimica Acta* 129:146–156.
Powis Terry G., Emiliano Gallaga Murrieta, Richard Lesure, Roberto Lopez Bravo, Louis Grivetti, Heidi Kucera, and Nilesh W. Gaikwad
2013 Prehispanic Use of Chili Peppers in Chiapas, Mexico. *PLOS ONE* 8(11):e79013. DOI:10.1371/journal.pone.0079013.
Rafferty, Sean M.
2002 Identification of Nicotine by Gas Chromatography/Mass Spectroscopy Analysis of Smoking Pipe Residue. *Journal of Archaeological Science* 29(8):897–907.
2016 Smoking Pipes of Eastern North America. In *Perspectives on the Archaeology of Pipes, Tobacco, and Other Smoke Plants in the Ancient Americas*, edited by E. Bollwerk and S. Tushingham, pp. 13–26. Springer, New York.
Rastrelli, Luca, Siro Passi, Ferdinando Ippolito, Giovanni Vacca, and Francesco De Simone
2002 Rate of Degradation of α-Tocopherol, Squalene, Phenolics, and Polyunsaturated Fatty Acids in Olive Oil during Different Storage Conditions. *Journal of Agricultural and Food Chemistry* 50(20):5566–5570. DOI:10.1021/jf011063j.

Raven, A. M., P. F. van Bergen, A. W. Stott, S. N. Dudd, and R. P. Evershed
1997 Formation of Long-Chain Ketones in Archaeological Pottery Vessels by Pyrolysis of Acyl Lipids. *Journal of Analytical and Applied Pyrolysis* 40–41:267–285.
Reber, Eleanora A.
2017 *Analysis of Ten Absorbed Residues from Hiwassee Island Pottery.* Papers of the UNCW Residue Lab 26. UNCW Anthropological Papers Vol. 35. UNC Wilmington, Wilmington, North Carolina.
2021 Comparison of Neutral Compound Extraction from Archaeological Residues in Pottery Using Two Methodologies: A Preliminary Study. *Separations* 8(1):6. DOI:10.3390/separations8010006.
Reber, Eleanora A., and Didi el-Behaedi
2016 In Their Cups?: Background Lipids in Shell as a Basis for Analyzing Shell Cup Residues. Paper presented at the 81st Annual Meeting of the Society for American Archaeology, Orlando, Florida.
Reber, Eleanora A., and Richard P. Evershed
2004 Identification of Maize in Absorbed Organic Residues: A Cautionary Tale. *Journal of Archaeological Science* 31(4):399–410.
Reber, Eleanora A., and John P. Hart
2008a Pine Resins and Pottery Sealing: Analysis of Absorbed and Visible Pottery Residues from Central New York State. *Archaeometry* 50(6):999–1017.
2008b Visible Clues: The Analysis of Visible Pottery Residues from New York State with Gas Chromatography/Mass Spectrometry. In *Current Northeast Paleoethnobotany II*, edited by John P. Hart, pp. 129–140. New York State Museum Bulletin Vol. 512. New York State Education Department, Albany.
Reber, Eleanora A., and Matthew T. Kerr
2012 The Persistence of Caffeine in Experimentally Produced Black Drink Residues. *Journal of Archaeological Science* 39(7):2312–2319.
Reber, Eleanora, John Blitz, and Claire Thompson
2010 Direct Determination of the Contents of a Ceramic Bottle from the Moundville Site, Alabama. *Midcontinental Journal of Archaeology* 35:37–55. DOI:10.1179/mca.2010.003.
Reber, Eleanora A., Timothy Baumann, William Monaghan, and Kelsey Noakes
2015 Absorbed Residue Analysis of a Mississippi Plain Jar from Angel Mounds (12Vg1): Lipid Distribution Revisited. *Advances in Archaeological Practice* 3(1):29–49.
Reber, Eleanora A., Stephanie N. Dudd, Nikolaas J. van der Merwe, and Richard P. Evershed
2004 Direct Detection of Maize in Pottery Residues via Compound Specific Stable Carbon Isotope Analysis. *Antiquity* 78(301):682–691.
Reber, Eleanora A., John E. Kelly, Elizabeth Boswell, and Chad S. Lane
2020 Molecular Evidence of Changing Foodways across the Mississippian Transition at the George Reeves site (11S650). *Southeastern Archaeology* 39(2):71–88.
Reber, Eleanora A., Matthew T. Kerr, Helen L. Whelton, and Richard P. Evershed
2019 Lipid Residues from Low-Fired Pottery. *Archaeometry* 61(1):131–144.
Regert, Martine, Helen A. Bland, Stephanie N. Dudd, Pim F. Van Bergen, and Richard P. Evershed

1998 Free and Bound Fatty Acid Oxidation Products in Archaeological Ceramic Vessels. *Proceedings of the Royal Society B* 265:2027–2032.

Ren, Meng, Zihua Tang, Xinhua Wu, Robert Spengler, Hongen Jiang, Yimin Yang, and Nicole Boivin

2019 The Origins of Cannabis Smoking: Chemical Residue Evidence from the First Millennium BCE in the Pamirs. *Science Advances* 5(6):eaaw1391. DOI:10.1126/sciadv.aaw1391

Rice, Prudence

2015 *Pottery Analysis: A Sourcebook*. 2nd ed. University of Chicago Press, Chicago.

Robinson, David W., Kelly Brown, Moira McMenemy, Lynn Dennany, Matthew J. Baker, Pamela Allan, Caroline Cartwright, Julienne Bernard, Fraser Sturt, Elena Kotoula, Christopher Jazwa, Kristina M. Gill, Patrick Randolph-Quinney, Thomas Ash, Clare Bedford, Devlin Gandy, Matthew Armstrong, James Miles, and David Haviland

2020 Datura Quids at Pinwheel Cave, California, Provide Unambiguous Confirmation of the Ingestion of Hallucinogens at a Rock Art Site. *Proceedings of the National Academy of Sciences* 117(49):31026–31037. DOI:10.1073/pnas.2014529117.

Robson, Harry K., Raminta Skipitytė, Giedrė Piličiauskienė, Alexandre Lucquin, Carl Heron, Oliver E. Craig, and Gytis Piličiauskas

2019 Diet, Cuisine, and Consumption Practices of the First Farmers in the Southeastern Baltic. *Archaeological and Anthropological Sciences* 11(8):4011–4024.

Roffet-Salque, Mélanie, Julie Dunne, David T. Altoft, Emmanuelle Casanova, Lucy J. E. Cramp, Jessica Smyth, Helen L. Whelton, and Richard P. Evershed

2017 From the Inside Out: Upscaling Organic Residue Analyses of Archaeological Ceramics. *Journal of Archaeological Science: Reports* (16):627–640.

Sathe, Shridhar, Erin Monaghan, Harshal Kshirsagar, and Mahesh Venkatachalam

2008 Chemical Composition of Edible Nut Seeds and Its Implications in Human Health. In *Tree Nuts: Composition, Phytochemicals, and Health Effects*, edited by Cesarettin Alasalvar and Fereidoon Shahidi, pp. 11–35. CRC Press, Boca Raton, Florida.

Schimmelmann, Arndt, Alex L. Sessions, and Maria Mastalerz

2006 Hydrogen Isotopic (D/H) Composition of Organic Matter during Diagenesis and Thermal Maturation. *Annual Review of Earth and Planetary Sciences* 34(1):501–533.

Seidel, Michael, Sarah Kleindienst, Thorsten Dittmar, Samantha Joye, and Patricia M. Medeiros

2016 Biodegradation of Crude Oil and Dispersants in Deep Seawater from the Gulf of Mexico: Insights from Ultra-High Resolution Mass Spectrometry. *Deep Sea Research II* 129:108–116.

Serpico, Margaret, and Raymond White

1996 A Report on the Analysis of the Contents of a Cache of Jars from the Tomb of Djer. In *Aspects of Early Egypt*, edited by J. Spencer, pp. 128–138. British Museum Press, London.

Sherratt, Andrew G.

1981 Plough and Pastoralism: Aspects of the Secondary Products Revolution. In *Pattern of the Past*, edited by Ian Hodder, Glyn Isaac, and Norman Hammond, pp. 261–306. Cambridge University Press, Cambridge.

Shoda, Shinya, Alexandre Lucquin, Chi Ian Sou, Yastami Nishida, Guoping Sun, Hiroshi Kitano, Joon-ho Son, Shinichi Nakamura, and Oliver E. Craig

2018 Molecular and Isotopic Evidence for the Processing of Starchy Plants in Early Neolithic Pottery from China. *Scientific Reports* 8:17044. DOI:10.1038/s41598-018-35227-4.

Simon, Mary L.

2002 Red Cedar, White Oak, and Bluestem Grass: The Colors of Mississippian Construction. *M.C.J.A.: Midcontinental Journal of Archaeology* 27(2):273–308.

Simpson, Ian A., Roland Bol, Ian D. Bull, Richard P. Evershed, Klaus J. Petzke, and Stephen J. Dockrill

1999 Interpreting Early Land Management through Compound Specific Stable Isotope Analyses of Archaeological Soils. *Rapid Communications in Mass Spectrometry* 13:1315–1319.

Sipponen, Arno

2013 *Coniferous Resin Salve, Ancient and Effective Treatment for Chronic Wounds—Laboratory and Clinical Studies.* Academic Dissertation. Department of Orthopedics and Traumatology, Helsinki University Hospital, University of Helsinki, Helsinki, Finland.

Spangenberg, Jorge E.

2016 Bulk C, H, O, and Fatty Acid Stable Isotope Analyses for Purity Assessment of Vegetable Oils from the Southern and Northern Hemispheres. *Rapid Communications in Mass Spectrometry* 30(23):2447–2461.

Spangenberg, Jorge E., and Nives Ogrinc

2001 Authentication of Vegetable Oils by Bulk and Molecular Carbon Isotope Analyses with Emphasis on Olive Oil and Pumpkin Seed Oil. *Journal of Agricultural and Food Chemistry* 49(3):1534–1540.

Spangenberg, Jorge E., Stefanie Jacomet, and Jörg Schibler

2006 Chemical Analyses of Organic Residues in Archaeological Pottery from Arbon Bleiche 3, Switzerland—Evidence for Dairying in the Late Neolithic. *Journal of Archaeological Science* 33(1):1–13.

Stacey, Rebecca J., C. R. Cartwright, and C. McEwan

2006 Chemical Characterization of Ancient Mesoamerican "Copal" Resins: Preliminary Results. *Archaeometry* 48:323–340.

Stacey, Rebecca J., Carl Heron, and Mark Q. Sutton

1998 The Chemistry, Archaeology, and Ethnography of a Native American Insect Resin. *Journal of California and Great Basin Anthropology* 20(1):53–71.

Steele, Valerie J., Ben Stern, and Andy W. Stott

2010 Olive Oil or Lard?: Distinguishing Plant Oils from Animal Fats in the Archeological Record of the Eastern Mediterranean Using Gas Chromatography/Combustion/Isotope Ratio Mass Spectrometry. *Rapid Communications in Mass Spectrometry* 24:3478–3484.

Stern, Ben, Carl Heron, Margaret Serpico, and Janine Bourriau

2000 A Comparison of Methods for Establishing Fatty Acid Concentration Gradients across Potsherds: A Case Study Using Late Bronze Age Canaanite Amphorae. *Archaeometry* 42:399–414.

Swanton, John R.
1979 *The Indians of the Southeastern United States*. Reprinted. Classics in Smithsonian Anthropology Vol. 2. Originally published 1946, Bulletin No. 137, Bureau of American Ethnology, Smithsonian Institution, Washington, DC.
Szpak, Paul, Jean-François Millaire, Christine D. White, and Fred J. Longstaffe
2012 Influence of Seabird Guano and Camelid Dung Fertilization on the Nitrogen Isotopic Composition of Field-Grown Maize (*Zea mays*). *Journal of Archaeological Science* 39(12):3721–3740. DOI:10.1016/j.jas.2012.06.035.
Taché, Karine, and Oliver E. Craig
2015 Cooperative Harvesting of Aquatic Resources Triggered the Beginning of Pottery Production in North-Eastern North America. *Antiquity* 343(90):177–190.
Taché, Karine, Manon Bondetti, Alexandre Lucquin, Marjolein Admiraal, and Oliver E. Craig
2019 Something Fishy in the Great Lakes? A Reappraisal of Early Pottery Use in North-Eastern North America. *Antiquity* 93(371):1339–1349.
Tarka, Stanley M., and Herbert H. Cornish
1982 The Toxicology of Cocoa and Methylxanthines: A Review of the Literature. *CRC Critical Reviews in Toxicology* 9(4):275–312.
Tulloch, A. P.
1974 Leaf Wax of *Portulaca oleracea*. *Lipids* 9(9):664–668.
1981 Composition of Epicuticular Waxes from 28 Genera of Gramineae: Differences between Subfamilies. *Canadian Journal of Botany* 59:1213–1221.
Tushingham, Shannon, Charles M. Snyder, Korey J. Brownstein, William J. Damitio, and David R. Gang
2018 Biomolecular Archaeology Reveals Ancient Origins of Indigenous Tobacco Smoking in North American Plateau. *PNAS* 115(46):11742–11747. DOI:10.1073/pnas.1813796115.
Urem-Kotsou, Dushka, Ben Stern, Carl Heron, and Kostas Kotsakis
2002 Birch-Bark Tar at Neolithic Makriyalos, Greece. *Antiquity* 76(294):962–967.
US Department of Agriculture, Agricultural Research Service
1992–2016 Dr. Duke's Phytochemical and Ethnobotanical Databases. http://phytochem.nal.usda.gov, DOI:10.15482/USDA.ADC/1239279.
2019 FoodData Central. http://fdc.nal.usda.gov.
Van der Merwe, Nikolaas J.
1982 Carbon Isotopes, Photosynthesis, and Archaeology. *American Scientist* 70:596–606.
Van der Merwe, Nikolaas J., and J. C. Vogel
1978 ^{13}C Content of Human Collagen as a Measure of Prehistoric Diet in Woodland North America. *Nature* 276:815–816.
Verhoeven, N. M., R. J. A. Wanders, B. T. Poll-The, J.-M. Saudubray, and C. Jakobs
1998 The Metabolism of Phytanic Acid and Pristanic Acid in Man: A Review. *Journal of Inheritable Metabolic Disease* 21:697–728. DOI:10.1023/A:1005476631419.
Vogel, J. C., and Nikolaas J. van der Merwe
1977 Isotopic Evidence for Early Maize Cultivation in New York State. *American Antiquity* 42(2):238–242.
Wang, Jiajing, Li Liu, Terry Ball, Linjie Yu, Yuanqing Li, and Fulai Xing

2016 Revealing a 5,000-Y-Old Beer Recipe in China. *Proceedings of the National Academy of Sciences* 113(23):6444–6448. DOI:10.1073/pnas.1601465113.

Washburn, Dorothy K., William N. Washburn, Petia A. Shipkova, and Mary Ann Pelleymounter

2014 Chemical Analysis of Cacao Residues in Archaeological Ceramics from North America: Considerations of Contamination, Sample Size, and Systematic Controls. *Journal of Archaeological Science* 50:191–207.

Whelton, Helen L., Mélanie Roffet-Salque, Kostas Kotsakis, Dushka Urem-Kotsou, and Richard P. Evershed

2018 Strong Bias towards Carcass Product Processing at Neolithic Settlements in Northern Greece Revealed through Absorbed Lipid Residues of Archaeological Pottery. *Quaternary International* 496:127–139. DOI:10.1016/j.quaint.2017.12.018.

Wilkinson, Robert E., and Herman S. Mayeux Jr.

1990 Composition of Epicuticular Wax on *Opuntia engelmannii*. *Botanical Gazette* 151(3):342–347.

Woodbury, Simon E., Richard P. Evershed, and J. Barry Rossell

1987 Purity Assessments of Major Vegetable Oils Based on $\delta^{13}C$ Values of Individual Fatty Acids. *Journal of American Oil Chemists Society* 75(3):371–379.

1998 $\delta^{13}C$ Analyses of Vegetable Oil Fatty Acid Components, Determined by Gas Chromatography-Combustion-Isotope Ratio Mass Spectrometry, after Saponification or Regiospecific Hydrolysis. *Journal of Chromatography A* 805(1–2):249–257.

Woodbury, Simon E., Richard P. Evershed, J. Barry Rossell, Rosemary E. Griffith, and Peter Farnell

1995 Detection of Vegetable Oil Adulteration Using Gas Chromatography Combustion/Isotope Ratio Mass Spectrometry. *Analytical Chemistry* 67:2685–2690.

Zagorevski, Dmitri V., and Jennifer A. Loughmiller-Newman

2012 The Detection of Nicotine in a Late Mayan Period Flask by Gas Chromatography and Liquid Chromatography Mass Spectrometry Methods. *Rapid Communications in Mass Spectrometry* 26:403–411. DOI:10.1002/rcm.5339.

INDEX

Page numbers in italics refer to figures or tables.
Page numbers in bold refer to glossary entries.

absorbed residues: definition, 8–9; formation, 9–11; preservation, *13*, 14–18, 40, 137, 158–61; sampling methods, 19–23, 66–67, 71–72
acidic methanol extraction, 21–22, *23*, 53, 55, 114, 143–46
acyl lipids: diacylglycerols, *15*, 17, **176**; monoacylglycerols, *15*, 17, **180**; triacylglycerols, *15*, 17, 102, 137, **184**
alcoholic beverages, 62, 80–81, 100–101, 166, *167*. *See also* beer; pulque; wine
algae, 35, 124, 132
alkaline methanol extraction, 21, 67, 72
alkaloids,15, 22, 36–37, 61–86, 106. *See also* datura; hallucinogens; methylxanthines; nicotine; tobacco
alkanes, *16*, 17, 21, 35, 38, 47, 138–39, 142
alkanoic acids (APHs), ω-(o-alkylphenyl), 37, 126, *130*, 133–34, **175**
alkanols, *16*, 17, 21, 38, 47, 56, 87, 92, 114, 117–19, 126, 138–39, 142
amaranth (*Amaranthus spp.*), 116, *121*
American Bottom, 52–56, 153–58
amyrin, 88, *89*, 92, 100
analysis processing time. *See* turnaround time
Angel Mounds, 129, 136–43, 159
Annedda, 93
appendix problem, 165
Apple Creek site, *154*
Arctic, 125
atropine. *See* datura
Audrey North site, *154*

Bayou des Familles site, *154*
Baytown culture, 45, *154*
beer, 80–1. *See also* alcoholic beverages
benzoins. *See* phenolic resins
betulin, 36, 88, 100–101
biomarker approach, 35–38
biomarkers, *15–16*, 19, 29, 35–38, 101. *See also* category biomarkers; isotopic biomarkers; process biomarkers; resource specific biomarkers
birch resin. *See* betulin
black drink. *See* methylxanthines; yaupon holly
blanks: freezer blank, 18, **178**; solvent or procedural blank, 57–58, 83, **183**
bound insoluble fraction, 22, 130, 145, **175**
broomcorn millet (*Paniceum miliaceum*), 100

Brunswick Town/Fort Anderson site (31BW376), 39–42
bug spray contamination. *See* insect repellent contamination
bulk stable isotope analysis, 110–13, 115, 120, 124

C_3 photosynthesis, 109
C_4 photosynthesis, 109
cacao, 72–73, 79, 100. *See also* methylxanthines
Cahokia (11S34/11MS2), 73–74, *154*, 156, 166
caffeine, *71*; alkaloid biomarker, 14, 36–37, 61–63, 65, 70–71, 78, 79, 85, 100, 166–67; analysis of, 28, *68–69*, 68–70; as a contaminant, 81–85; extraction of, 22, 66, 72–73; survival in residues, 74–78. *See also* methylxanthines
calcium oxalate, 80–81
calibration of compound-specific isotope mass spectrometer, 113–14
cannabinoids, 61, 102
cannabis (*Cannabis spp.*), 86, 102
capsaicin, 61, 79
carbon cycle, 111–12; marine, 111, 124–25; terrestrial, 107–9. *See also* photosynthesis; trophic shift
carbon stable isotope analysis, 106–10, *107*
category biomarkers, 35–37, 40, 47, 85–86, 90, 100
chili peppers. *See* capsaicin
chloroform/methanol extraction, 21–22, *23*, 39, 66–67, 72, 78, *130*, 130, 135, 144–45
chocolate. *See* cacao
cholesterol: carbon isotope values of, 124, 126; as a category biomarker, 15, 35; as a potential contaminant, 58; specific interpretations of, 92, 122, *130*, 136, 138, 142–43
chromatography, 25–30
cinnamic acid, 101, *102*
citric acid, 80
cocaine, 61
coffee, 71, 84–85. *See also* caffeine; methylxanthines
Coles Creek period, 117, *154*, 155–56
colonoware, 39–41, 166
common fatty acids, 32, 35, 70, 102, 115
communic acid, 88–90
communication with analyst, 61–62, 85, 96, 104, 169–72
compound-specific isotope analysis (CSIA), 105–6, 111–15, 128, 172, **176**; dairy products, 126–28, 136, 162–64; detection of C_3 photosynthesis, 109; detection of C_4 photosynthesis, 115–18, 120, 153; effects of limestone bedrock on fish/shellfish lipids, 131–32; maize, 109, 112, 115–18, 152–58, 172; marine carbon, 124–26, 129, 131; ruminant fat, 126–28, 136
conifer resins, 36, 42, 87–99, 166, *167*. *See also* diterpenoids
contamination, *24*, 44; alkaloids, 81–85; crude oil, 45–49, 101; depositional, 35, 44–49, 80; instrument, 29, 114; laboratory, 57–58, 80; lotions, 51–52, 140; museum storage/legacy collections, 56–57; postdepositional, 49–50, 171, 173; solvent or procedural blanks, 57–58. *See also* fingerprint contamination; insect repellent contamination; labeling contamination; plasticizer contamination; sunscreen contamination
controls. *See* blanks
cooking methods, 11, 37, 134
copal, 87, 90, 100
Corexit. *See* dispersants
corn (*Zea mays*). *See* maize
cost of analysis, 163, 170–71
crassulacean acid metabolism (CAM) photosynthesis, 109, **175–76**

dahoon holly (*Ilex cassina*), 71, 73
dairy products/dairying, 111, 126–28, 136, 158, 162–64, 166, *168*, 172
dammar gum, 100–101
dark gray literature, 164

datura (*Datura stramonium*), 29, 63–64, *64*, 78–79
DDT, 49
Deepwater Horizon oil spill (MC252), 45–49, 100. *See also* dispersants
dehydroabietic acid, *89*, 91–92, 96–97, *141*
derivatization, 79, 113–14, **177**. *See also* gas chromatography
diagenesis, 34, **177**. *See also* hydrolysis; microbial degradation/breakdown; oxidation
dispersants, 45–48, 59
diterpenoids, *16*, 36, 87–97, 104, 138

eastern redcedar (*Juniperus virginiana*), 36, 97–98, *97*
Emergent Mississippian, 98, *154*, 155
Emerson site, *154*
empty vessels. *See* insufficient residue for interpretation
encrustations. *See* visible residues
Evie site, *154*
experimental residue production, 12, 17–18, 73–78, 94–95, 100, 111, 134
external spike, 70, **178**
extracted ion chromatograms (EIC), 67–68, **177**
extraction methodologies. *See* acidic methanol extraction; alkaline methanol extraction; chloroform/methanol extraction; hot water extraction; targeted extraction

fatty acid abundance interpretation, 19, 38–39, 47, 143–44
fatty acid composition, 13, 38, 49, 53, 119, 122, 132–36
fatty acids. *See* alkanoic acids (APHs), ω-(o-alkylphenyl); common fatty acids; isoprenoid fatty acids; monounsaturated fatty acids; polyoxyethylene (POE) fatty acid esters; polyunsaturated fatty acids; saturated fatty acids; unsaturated fatty acids
faunal analysis, 165–66, 169
fingerprint contamination, 20, 57–59, 136
fish: anadromous, 115, 124, *125*; extraction methodologies, 143–46; masking, 129, *130*, 131, 136. *See also* marine fish/shellfish; freshwater fish/shellfish
fishing expedition. *See* preliminary project
foodcrust. *See* visible residues
freshwater fish/shellfish: indicators, 129–36; isotopic values, 115, 131–32; interpretation, 125, 136–43, 146
freshwater reservoir effect (FRE), 131–32, **178**
Fourier transform infrared spectroscopy (FTIR), 24–25, *24*, **178**
functional group, 148, 151–52, *154*, 155, 159, 161–64, *167–68*
fungus, *15*, 35, 80

gas chromatography (GC), *24*, 25–31, 33, 45, 70, 105, 110, 112, **178**. *See also* gas chromatography-isotope ratio mass spectrometry (GC-IRMS); gas chromatography/mass spectrometry (GC/MS)
gas chromatography-isotope ratio mass spectrometry (GC-IRMS), 112–14, 171–72, **179**. *See also* compound-specific isotope analysis (CSIA)
gas chromatography/mass spectrometry (GC/MS), *27*, 30–31, 44, 66–68, 77–79, 112, **179**
George Reeves site (11S650), 52–56, 96–97, 131, 143–45
Gottschall rockshelter, *154*, 155, 158
guarana (*Paullinia cupana*), 36, 71
Gullah, 42

Halliday site (11S62), *154*, 156
hallucinogens, 61, 78. *See also* alkaloids; psychoactive substances
Hill Creek site, *154*
Hopanoids, 40, 42, 45, 47, 81, 100–101
hot water extraction, 22, 67, 72, 78, *83*, 84
high performance liquid chromatography/mass spectrometry (HPLC/MS), *24*, 28–33, 44, 67, 70, 72, 78–79, 83, 171–72

hydrogen stable isotope analysis, 106, *107*
hydrolysis, 14–18, 52, 77, 90, 94, 102, 104, 144, 146, **179**

Illinois River valley, *154*, 154, 158
indicators: alcoholic beverage detection, 80; animal products, 35; C_4 plant detection, 118; conifer resin, 36; contamination detection, 58–59; development, 171–72; fish/shellfish detection, 37, 42, 129–30, 132–37, 141–43, 145–46; ku-nu-che processing, 103; overall vessel contents, 38; phenolic resin detection, 101; marine fats, 125
infrared spectroscopy. *See* Fourier transform infrared spectroscopy (FTIR)
insect-produced resins, *16*, 87, 118
insect-repellant contamination, 49, 51, 53, 141
insect wax, *16*, 87, 118,
insufficient residue for interpretation, 42, 144, 158
internal standard, 14, 42, 57, **180**
interpretable residues, yield of, 158–60
isopimaric acid, 88, *89*, *141*
isoprenoid fatty acids, 22, *130*, 134–35, 142–45
isotopic biomarkers, 116, 118
isotopic fractionation, 126. *See also* trophic shift
isotopic masking, 118–20, 122–24, 156
isotopic mixing models, 122–23

Jersey Bluff phase, *154*, 155

Kalinga, 93
ketones, 37, 39, 42
Kipp Island site, *92*, 97
kola nut, 71. *See also* methylxanthines
Koster East site, *154*
ku-nu-che, 103

labeling contamination, 50–56, 60
lac resins. *See* insect-produced resins
lactic acid, 80
Late Woodland period, *154*, 155
Leguminosae, 90
levoglucosan, 161
lipid composition. *See* fatty acid composition
lipid levels, 120, *121–22*
liquid chromatography, *24*, 25, 28–29, **180**. *See also* high performance liquid chromatography/mass spectrometry (HPLC/MS)
Lower Mississippi Valley, 153, *154*, 155–58

maize (*Zea mays*), 90, *168*; bulk isotopic detection, 111; CSIA detection, 112, 115–18, 152–58, 172; masking, 119–24, 160; residue formation and preservation, *13*, 14, 17; stable isotopic identification, 107, 109
malic acid, 80
marijuana. *See* cannabis and cannabinoids
marine carbon. *See* compound-specific isotope analysis, marine carbon
marine fish/shellfish, 106, 124–26, 129, 131
masking of lipid poor resources, 7, 118–20, 122–24, 129, *130*, 131, 136, 156, 161, *167*
mass spectrometry, *24*, 26, 30–34, **181**; isotope ratio, 112, *113*; MS/MS, 33, **181**; quadrupole, 33, 68–70, **182**; time of flight (TOF), 33, 70, **184**
Mees-Nochta site (11S1198), *154*, 156
mescaline, 61
methylxanthines, 70–80, 82–84. *See also* caffeine; theobromine; theophylline
microbial degradation/breakdown, 14, 16–18, 77, 126–17, *130*, 136, 138, 144, 146, **181**
miliacin, 100–101
minimizing variables, 148, 150–51, 160–61
mixed residues, 10, 12, 35, 38, 74, 105, 119–20, 122–23, 128, 136–37, 143–44, 146, 161, *167–68*. *See also* masking of lipid poor resources; residue formation
Mississippi Plain pottery, 137–43
Mississippi river, 153–54

Mississippian period, 52, 78–79, 98, 137, 152, *154*, 156
mixing models. *See* isotopic mixing models
monoterpenoids, 87
monounsaturated fatty acids, *130*, 132, 136, 142–43, 145
Morgan Mounds, *154*

n-dotriacontanol (OL_{32}), 117–19, 155, *157*
negative painted pottery (NPP), 137, 142–43
neutral compounds. *See* alkanes; alkanols; sterols; terpenoids
nicotine, 7, 14, 36, 61–63, 78–82
nitrogen stable isotope analysis, 106, *107*, 112
nonconiferous plant resins, 36, 88, 90, 92, 99–100, 167
nuts, 90, 102–3, 119–23, 129, 160, 171

olive oil, 4, 38, 59, 65, 112
opium, 63
Osceola site, 117, *154*
outside funding, 172–73
oxidation, 5, 77, **181**; fatty acids, 14–17, 37, 52, 104, 132, 146; terpenoids, 89, 91, 94; phenolic resins, 102; squalene, 58–59. *See also* diagenesis
oxygen stable isotope analysis, 106, *107*

paleoethnobotany, 43, 87, 104, 117, 165–66, 169
Pamunkey tribe, 93
panicgrass (*Panicum spp.*), 116, 118
petroleum, 35, 40, 42, 45
phenolic resins, 61, 86–87, 98–99, 101–2, 167
photosynthesis, 105–8, 111, 124. *See also* C_3 photosynthesis; C_4 photosynthesis; crassulacean acid metabolism (CAM) photosynthesis
phytoliths, 81
pipes, 63, 78–79, 85, *167*. *See also* nicotine
plant resins, 86–102, 104, 138
plant wax, 37, 118, 138
Plaquemine Mississippian period, *154*, 156
plasticizer contamination, 25, 50, 53–56, *139*, 143, 169
polyaromatic hydrocarbons (PAHs), 95, **181**
polyoxyethylene (POE) fatty acid esters, 47–48, **182**
polyunsaturated fatty acids, 37, *130*, 132–33, 135, **182**
pottery sealing and/or sealants, 90, 93–95, 98–99, 101, 104, *167–68*
preliminary project, 148–51, 159, 164, 173
prickly pear cactus (*Opuntia spp.*), 116, 118
process biomarkers, 37, 40, 92, 100, 133–34
project sample size, 147, 153, 155, 160–64. *See also* functional group; minimizing variables; sampling strategies
psilocybin, 61, 63
psychoactive substances, 61–66, 80–81, 85–86, 102, 104, 166, *167*. *See also* alkaloids
pulque, 81, 100–101
purslane (*Portulaca sp.*), 116, 118
pyrolytic reactions, 37, 40. *See also* process biomarkers

quantification, 29, 42, 55, **183**

Redfish Slough site, *48*
Reno Brake site, *154*
research design, 85, 151, 162
research questions, 148–52, 158, 161, 163–64, *167–68*, 171–72
reservoir effect. *See* freshwater reservoir effect
residue breakdown. *See* diagenesis
residue formation, 14–15; absorbed residues, 9–12, 14; visible residues, 11–12, 16
residue interpretation: contamination and interpretability, 50–52, 55–56, 60. *See also* biomarker approach; fatty acid abundance interpretation
residue preservation, 16, 159–60. *See also* diagenesis; hydrolysis; microbial degradation/breakdown; oxidation

resin. *See* conifer resins; diterpenoids; non-coniferous plant resins; insect produced resins; Leguminosae; plant resins; triterpenoids
resinous plants, 36, 88, 98–101, 104
resource mixtures. *See* mixed residues
resource specific biomarkers, 36–37
retene, 92
rice (*Oryza sativa*), 161
Richland Complex, 156
ricinoleic acid, 56
ruminant fat, 105, 112, 126–28, 136. *See also* compound-specific isotope residue analysis (CSIA)

sampling methods: burr sampling, 20–21, 79; destructive, 19–20, 63, 66, 147, 159, 173; non-destructive, 19, 78, 83. *See also* absorbed residues; visible residues
sampling strategies: Bayesian, 151; experience-based, 147; interpretable residues, 151–52, *154*, 158–59, 162, 164, *167*; statistical, 147–48; uninterpretable residues, 158–59, 162, *167*
saturated fatty acids, *15*, 17, 37–38, 52, 132, 135, **183**
scan mode (MS), 68–70
scopolamine. *See* datura
secondary products revolution, 162–63
selected ion monitoring (SIM) mode, 68–70, 78, 100, **183**
shelf swabbing, 56–57, 82. *See also* contamination
shell cups, 63–65, 79–80, 82–84, *83*
sitosterol: category biomarker, *15*, 35, 40, 122; CSIA analysis of, 116–17
smoking pipes. *See* pipes
soil chemistry, 18, 44, 73, 159
solvent extraction. *See* chloroform/methanol extraction
soot, 88, 94–95, 181
Sorbitan, 80, 47
Southern Comfort site, *46*
spectrometry. *See* mass spectrometry
squalene, 58–59, 141
stable isotope analysis, 105–29, *107*, 131–32. *See also* bulk stable isotope analysis; carbon stable isotope analysis; compound-specific stable isotope analysis (CSIA), hydrogen stable isotope analysis; nitrogen stable isotope analysis; oxygen stable isotope analysis
starch grain analysis, 81, 166
starchy grains, 129, 161
Starr Village, *154*
sterols, 21, 26, 88; category biomarker, *15*, 35–38, 47, 50, 86; contamination, 51–52; CSIA analysis, 114, 126, 131; plant sterols, 35–36, 40, 116–17, 138–39, 142. *See also* biomarkers; cholesterol; sitosterol
St. Gabriel Mound, *154*
stone pipes (smoking), 79. *See also* nicotine
succinic acid, 80
sunscreen contamination, 49, 51–53, 58, 141
sweetgum (*Liquidambar styraciflua*), 86, 101–102
syringic acid, 80

targeted extraction, 6–7, 66–67, 72–73, 78, 166
tartaric acid, 80
tea (*Camellia sinensis*), 29, 36, 71–72. *See also* methylxanthines
Terminal Late Woodland. *See* Emergent Mississippian
terpenoids. *See* diterpenoids; hopanoids; monoterpenoids; triterpenoids
theobromine, 29, 62, 676–78, 71–74, 78. *See also* methylxanthines
theophylline, 62, 66, 71–72. *See also* methylxanthines
Tianluoshan site, 161
tobacco. *See* nicotine
triterpenoids, *16*, 36, 81, 86–90, 92, 99–104
trophic shift, 106–8, *107*, 110–12, 115, **185**
turnaround time, 170–71
turpentine, 91

uninterpretable residues: contamination, 50–52, 56–57, 60; insufficient residue for interpretation, 158, *167*
unsaturated fatty acids: compound-specific isotopic analysis, 126–28; definition, *16*; indicators for fish/shellfish, 129–30, 132–34, 136, 138, 142–45; interpretation, 38–39, 41, 51–52, 58–59, 122; oxidation, 17, 37–38; preservation, 40, 145; pyrolytic reactions, 37. *See also* monounsaturated fatty acids; polyunsaturated fatty acids
ursolic acid, 73, 100

Vinette site, 91
Vinette pottery, 125
visible residues: analysis, 25, 90–91, 101, 161–62; bulk isotopic analysis, 110–11; calcium oxalate, 81; definition, 8; extraction, 19; formation, 11–12, 16; nitrogen stable isotope analysis, 106; sampling, 63, 78, 169

Walsh site, *154*
waxes, *16*. *See also* insect wax; plant wax
wine, 65, 80, 168

yaupon holly (*Ilex vomitoria*), 36, 63, *65*, 71–77, *74*, 100. *See also* methylxanthines
yerba mate (*Ilex paraguariensis*), 36, 71. *See also* methylxanthines